Time-Fractional Order Biological Systems with Uncertain Parameters

Synthesis Lectures on Mathematics and Statistics

Editor
Steven G. Krantz, *Washington University, St. Louis*

Time-Fractional Order Biological Systems with Uncertain Parameters
Snehashish Chakraverty, Rajarama Mohan Jena, and Subrat Kumar Jena

ISBN: 978-3-031-01295-2   paperback
ISBN: 978-3-031-02423-8   ebook
ISBN: 978-3-031-00269-4   hardcover

DOI 10.1007/978-3-031-02423-8

A Publication in the Springer series
SYNTHESIS LECTURES ON MATHEMATICS AND STATISTICS

Lecture #31
Series Editor: Steven G. Krantz, *Washington University, St. Louis*
Series ISSN
Print 1938-1743   Electronic 1938-1751

Time-Fractional Order Biological Systems with Uncertain Parameters

Snehashish Chakraverty, Rajarama Mohan Jena, and Subrat Kumar Jena
National Institute of Technology Rourkela

SYNTHESIS LECTURES ON MATHEMATICS AND STATISTICS #31

ABSTRACT

The subject of fractional calculus has gained considerable popularity and importance during the past three decades, mainly due to its validated applications in various fields of science and engineering. It is a generalization of ordinary differentiation and integration to arbitrary (non-integer) order. The fractional derivative has been used in various physical problems, such as frequency-dependent damping behavior of structures, biological systems, motion of a plate in a Newtonian fluid, $PI^{\lambda}D^{\mu}$ controller for the control of dynamical systems, and so on. It is challenging to obtain the solution (both analytical and numerical) of related nonlinear partial differential equations of fractional order. So for the last few decades, a great deal of attention has been directed towards the solution for these kind of problems. Different methods have been developed by other researchers to analyze the above problems with respect to crisp (exact) parameters.

However, in real-life applications such as for biological problems, it is not always possible to get exact values of the associated parameters due to errors in measurements/experiments, observations, and many other errors. Therefore, the associated parameters and variables may be considered uncertain. Here, the uncertainties are considered interval/fuzzy. Therefore, the development of appropriate efficient methods and their use in solving the mentioned uncertain problems are the recent challenge.

In view of the above, this book is a new attempt to rigorously present a variety of fuzzy (and interval) time-fractional dynamical models with respect to different biological systems using computationally efficient method. The authors believe this book will be helpful to undergraduates, graduates, researchers, industry, faculties, and others throughout the globe.

KEYWORDS

fractional calculus, fuzzy set theory, time-fractional differential equations, mathematical models

Contents

Preface

Real-life problems, such as biological disease models, may be well described using ordinary and partial differential equations. Usually, these phenomena are modeled by ordinary or partial differential equations with integer order. On the other hand, a non-integer order derivative makes it possible to measure the correct information at each instant of time. Further, it may be difficult to express the dynamics between two different points when the derivative is in the integer order. Moreover, the fractional derivatives have a description of memory and heredity properties. As such, fractional calculus may be well suited for real-life problems. Thus for the biological disease models, the fractional derivative may give more useful information than classical derivatives. The subject of fractional calculus has gained considerable popularity and importance during the past three decades, mainly due to its validated applications in various fields of science and engineering. It deals with the differential and integral operators with non-integral powers. Fractional differential equations are the pillar of various systems occurring in a wide range of science and engineering disciplines, namely, physics, chemical engineering, mathematical biology, financial mathematics, structural mechanics, control theory, circuit analysis, biomechanics, and so on. Moreover, every physical problem is inherently biased by uncertainty. In general, parameters, variables, and initial conditions involved in the model are considered crisp or exactly defined for easy computation. However, rather than the particular value, we may have only the vague, imprecise, and incomplete information about the variables and parameters as a result of errors in measurement, observations, experiment, or applying different operating conditions, or there may be maintenance-induced errors, which are uncertain in nature. So, in order to understand these uncertainties and vagueness, one may use either a stochastic/statistical approach or interval/fuzzy set theory. But stochastic and statistical uncertainty occurs due to the natural randomness in the process. It is generally expressed by a probability density or frequency distribution function. For the estimation of the distribution, it requires sufficient data about the variables and parameters involved in it. On the other hand, interval and fuzzy set theory refer to the uncertainty when we may have a lack of knowledge or incomplete information about the variables and parameters. As such, in this book, the fuzzy set theory has been used for understanding different well-known biological disease models defined in an uncertain environment. These uncertainties are introduced in the fractional-order differential equations defined in the Caputo sense. Due to the complexity of the fuzzy arithmetic, one may need reliable and efficient analytical and numerical techniques for the solution of fuzzy fractional-order differential equations.

In view of the above, this book starts with the preliminaries of fractional-order differential equations along with the basics of fuzzy and interval theory. Then, a computationally efficient method, namely, fractional reduced differential transform method (FRDTM), is applied to solve various fractional-order biological disease models in uncertain environments. As such, this book consists of eight chapters. The authors assume readers have essential knowledge of calculus, differential equations, fractional calculus, functional analysis, real analysis, and linear algebra.

Accordingly, Chapter 1 addresses the preliminaries of fractional calculus in which various important functions related to the fractional calculus and popular differential and integral operators of fractional order have been included. Basics of fuzzy set theory and fuzzy fractional initial and boundary value problems are presented in Chapter 2. In Chapter 3, a computationally efficient method viz. FRDTM is discussed for solving fractional and fuzzy fractional differential equations. In Chapters 4 and 5, investigation of imprecisely defined time-fractional model of cancer chemotherapy effect and fuzzy time-fractional smoking epidemic model are incorporated, respectively. In these two models, fuzziness in the involved parameters are considered in terms of triangular fuzzy number. Chapters 6, 7, and 8 address different models of diseases using time-fractional differential equations in the uncertain environment with respect to HIV infection of CD4+ T Lymphocyte cells, hepatitis E virus, and the SIRS-SI malaria disease model, respectively. In these models, fuzziness in the initial conditions have been taken in terms of triangular fuzzy numbers.

This book is an attempt to rigorously address the challenge of a variety of uncertain (fuzzy and interval) time-fractional dynamical models with respect to biological systems using the computationally efficient method. The authors believe this book will be helpful to undergraduates, graduates, researchers, industry, faculties, and others throughout the globe.

Snehashish Chakraverty, Rajarama Mohan Jena, and Subrat Kumar Jena
January 2020

Acknowledgments

Writing this book was an amazing journey that would not have been possible without the continuous support, encouragement, and motivation received from different outstanding people who not only guided us through our hardships but also made us believe that we could achieve what we wanted.

As such, the first author would like to thank his parents for their blessings and motivation. Next, he would like to thank his wife, Shewli, and daughters, Shreyati and Susprihaa, for their support and being a source of inspiration during this project. The support of all the Ph.D. students of the first author as well as the NIT Rourkela facilities is also gratefully acknowledged.

The second and third authors would like to express their sincere gratitude to their family members, especially Sh. Ullash Chandra Jena, Sh. Durga Prasad Jena, Sh. Laxmidhara Jena, Smt. Urbasi Jena, Smt. Renu Bala Jena, Smt. Arati Jena, and sisters Jyotrimayee, Truptimayee, and Nirupama for their enormous love, continuous motivation, support, and blessings. Further, the second author would like to acknowledge the Department of Science and Technology, Government of India, for providing INSPIRE fellowship (IF170207) to carry out the present work. Last but not least, the second author also wants to thank Miss Sujata Swain for showing her continuous love, support, encouragement, and faith.

Also, the second and third authors greatly appreciate the inspiration of the first author for his support and inspiration. This work would not have been possible without his guidance, support, and encouragement.

Further, the authors sincerely acknowledge the reviewers for their fruitful suggestions and appreciations. They also appreciate the support and help of the whole team of Morgan & Claypool Publishers. Finally, the authors are greatly indebted to all the authors/researchers mentioned in the reference sections given at the end of each chapter.

Snehashish Chakraverty, Rajarama Mohan Jena, and Subrat Kumar Jena
January 2020

CHAPTER 1

Preliminaries to Fractional Calculus

1.1 INTRODUCTION

In fractional calculus, differentiation and integration are of arbitrary or non-integer order. This area is three centuries old compared to conventional calculus, but initially, it was not very popular. Fractional derivatives and integrals are not local in nature, and so the non-local distributed effects are considered. The subject of fractional calculus has gained considerable popularity and importance during the past three decades, mainly due to its validated applications in various fields of science and engineering. The mathematical models in electromagnetics, rheology, viscoelasticity, electrochemistry, control theory, fluid dynamics, financial mathematics, and material science are well defined by fractional-order differential equations.

1.2 BIRTH OF FRACTIONAL CALCULUS

In a letter to L'Hospital in 1695, Leibniz asked the following question: "***Can the meaning of integer-order derivatives be generalized to non-integer order derivatives***?" L'Hospital was very much curious about that question and replied to Leibniz by asking another question of what will happen to the term $\dfrac{d^n \psi(x)}{dx^n}$ if $n = \frac{1}{2}$. In order to explain the answer to the query raised by L'Hospital, Leibniz wrote a letter dated September 30, 1695, known as the birthday of fractional calculus, saying, "***It will lead to a paradox, from which one-day useful consequences will be drawn.***" This was the beginning of fractional calculus. Many famous mathematicians, including Liouville, Reimann, Weyl, Fourier, Abel, Lacroix, Leibniz, Grunwald, and Letnikov, and others, contributed to fractional calculus over the years.

There are various ways of defining fractional derivatives and integrals. A few are incorporated below.

1.3   POPULAR DEFINITIONS OF FRACTIONAL DERIVATIVES AND INTEGRALS

In order to understand the important definition of fractional derivatives and integrals in fractional calculus, we need first to understand some necessary preliminaries related to fractional calculus.

Definition 1.1   Gamma function.   The gamma function is most important in fractional-order calculus, and it is written as

$$\Gamma(n) = \int_0^\infty e^{-x} x^{n-1} dx. \tag{1.1}$$

Some properties of the gamma function are as follows:

(i) $\Gamma(n+1) = n\Gamma(n)$, for $n \in \Re^+$.

(ii) $\Gamma(n+1) = n!$, $\Gamma(1) = 1$ and $\Gamma\left(\frac{1}{2}\right) = \sqrt{\pi}$.

(iii) $\Gamma\left(\frac{1}{2} - n\right) = \frac{n!(-4)^n}{(2n)!}\sqrt{\pi}$ and $\Gamma\left(\frac{1}{2} + n\right) = \frac{(2n)!}{(4)^n n!}\sqrt{\pi}$.

(iv) $\Gamma(x)\Gamma(-x) = \frac{-\pi}{x\sin(\pi x)}$.

Definition 1.2   One-parametric Mittag–Leffler function (MLF).   One-parameter MLF is defined as

$$E_\alpha(z) = \sum_{n=0}^\infty \frac{z^n}{\Gamma(1 + n\alpha)}, \quad \text{for} \quad z \in C \quad \text{and} \quad \alpha > 0, \tag{1.2}$$

where C represents the complex number.

If we put $\alpha = 1$ in Eq. (1.2), we obtain

$$E_1(z) = \sum_{n=0}^\infty \frac{z^n}{\Gamma(1 + n)}, \quad \text{for} \quad z \in C \tag{1.3}$$

which is the summation form of exponential function e^z. So, MLF is an extension of the exponential function in one parameter.

Definition 1.3   Two-parametric MLF.   Two-parameter representation of the MLF may be written as

$$E_{\alpha,\beta}(z) = \sum_{n=0}^\infty \frac{z^n}{\Gamma(\beta + n\alpha)}, \quad \text{for} \quad z \in C \quad \text{and} \quad \alpha, \beta > 0. \tag{1.4}$$

Definition 1.4 Generalized MLF.   In general, the MLF is written as

$$E_{\alpha,\beta}^{\gamma}(z) = \sum_{k=0}^{\infty} \frac{(\gamma)_n}{\Gamma(\beta + k\alpha)} \frac{z^k}{k!} \quad \text{for} \quad z, \alpha, \beta \quad \text{and} \quad \gamma \in C, \tag{1.5}$$

where

$$(\gamma)_n = \frac{\Gamma(\gamma + n)}{\Gamma(\gamma)} = \begin{cases} 1, & n = 0, \quad \gamma \neq 0, \\ (\gamma + n - 1) \cdots (\gamma + 2)(\gamma + 1)\gamma, & n \in N, \quad \gamma \in C. \end{cases}$$

The term N denotes the natural number.

There are various ways of defining fractional derivatives. As such, preliminaries of some definitions are incorporated below for the sake of completeness.

Definition 1.5 Riemann–Liouville (R-L) fractional integral operator.   The Riemann–Liouville (R-L) fractional integral operator J_t^{α} of an order α of a function $\psi(t)$ is defined as:

$$J_t^{\alpha}\psi(t) = \frac{1}{\Gamma(\alpha)} \int_0^t (t - \xi)^{\alpha - 1} \psi(\xi)d\xi, \quad t > 0 \quad \text{and} \quad \alpha \in R^+. \tag{1.6}$$

Proof. Let us take the first integral of a function $\psi(t)$ as

$$J_t^1 \psi(t) = \int_0^t \psi(u)\,du, \quad t > 0. \tag{1.7}$$

Integrating Eq. (1.7) once more, we have

$$J_t^2 \psi(t) = \int_0^t \int_0^u \psi(v)\,dv\,du, \quad t > 0. \tag{1.8}$$

Successive integration of $\psi(t)$ for n times (n, integer) is

$$J_t^n \psi(t) = \int_0^t \int_0^u \cdots \int_0^w \psi(v)\,dv\,dw \ldots du, \quad t > 0. \tag{1.9}$$

The closed-form formula of Eq. (1.9) which is given by Cauchy is as follows:

$$J_t^n \psi(t) = \frac{1}{(n-1)!} \int_0^t (t - u)^{n-1} \psi(u)\,du, \quad t > 0. \tag{1.10}$$

Now, just by replacing n by α and factorial by the gamma function of Eq. (1.10), we may have the desired result (1.6). $\square$

Definition 1.6   R-L left, and right integral operator.   The left- and right-hand sides R-L fractional integral of a function $\psi(t)$ are defined, respectively, as

$$_{-\infty}J_t^\alpha \psi(t) = \frac{1}{\Gamma(m-\alpha)} \int_{-\infty}^{t} (t-\xi)^{m-\alpha-1} \psi(\xi)\, d\xi, \quad \text{for} \quad m-1 < \alpha < m, \quad m \in N$$

(1.11)

and

$$_tJ_\infty^\alpha \psi(t) = \frac{(-1)^m}{\Gamma(m-\alpha)} \int_{t}^{\infty} (\xi-t)^{m-\alpha-1} \psi(\xi)\, d\xi, \quad \text{for} \quad m-1 < \alpha < m, \quad m \in N.$$

(1.12)

Proof. Hints: Substituting $m-\alpha$ in place of α in Eq. (1.6) and taking integration over $-\infty$ to t, we get left R-L integral. Similarly, right R-L integral is obtained using the same technique, but the $(-1)^m$ term arises due to change of the order of integration $\left(\text{i.e., } \int_a^b f(x)\, dx = -\int_b^a f(x)\, dx\right)$. One may see the detail derivations of the above definitions in any standard fractional calculus book for this concept.

From Podlubny [6], we have the following result:

$$J_t^\alpha t^n = \frac{\Gamma(n+1)}{\Gamma(n+\alpha+1)} t^{n+\alpha}, \quad n > -1, \quad \alpha > -1-n.$$

(1.13)

□

Proof. We know that mth time integration of t^n is defined as

$$J_t^m t^n = \frac{\Gamma(n+1)}{\Gamma(n+m+1)} t^{n+m}, \quad n > -1.$$

(1.14)

In place of $m-\alpha$, if we substitute α in Eq. (1.14), then we obtain α times integration of t^n, which is defined as in Eq. (1.13).

□

Example 1.7   Suppose we want to calculate $\frac{1}{2}$ times integration of $\psi(t) = t$; then we have

$$J_t^{0.5} t = \frac{\Gamma(1+1)}{(1+0.5+1)} t^{1+0.5} = \frac{1}{\Gamma(2.5)} t^{1.5} = \frac{3}{4}\sqrt{\pi} t^{1.5}.$$

(1.15)

Definition 1.8 R-L fractional differentiation operator. The fractional-order R-L derivative of order α is defined as

$$
D_t^\alpha \psi\,(t) = \begin{cases} \dfrac{1}{\Gamma\,(m-\alpha)} \dfrac{d^m}{dt^m} \displaystyle\int_0^t (t-\xi)^{m-\alpha-1} \psi\,(\xi)\, d\xi, & \text{for} \quad m-1 < \alpha < m, \quad m \in N, \\[3mm] \dfrac{d^m}{dt^m} \psi\,(t), & \text{for} \quad \alpha = m, \quad m \in N. \end{cases}
$$

$$(1.16)$$

Proof. Hints: we know that

$$
\frac{d^\alpha \psi\,(t)}{dt^\alpha} = D_t^\alpha \psi\,(t) = \left(D_t^m D_t^{\alpha-m} \right) \psi\,(t)\,.
$$

Or,

$$
D_t^\alpha \psi\,(t) = \left(D_t^m D_t^{-(m-\alpha)} \right) \psi\,(t) = \left(D_t^m J_t^{(m-\alpha)} \right) \psi\,(t)\,. \tag{1.17}
$$

Now, using Eq. (1.6), Eq. (1.17) yields

$$
D_t^\alpha \psi\,(t) = D_t^m \frac{1}{\Gamma\,(m-\alpha)} \int_0^t (t-\xi)^{m-\alpha-1}\, \psi\,(\xi)\, d\xi.
$$

Or,

$$
D_t^\alpha \psi\,(t) = \frac{1}{\Gamma\,(m-\alpha)} \frac{d^m}{dt^m} \int_0^t (t-\xi)^{m-\alpha-1}\, \psi\,(\xi)\, d\xi. \tag{1.18}
$$

□

Definition 1.9 Riemann–Liouville (R-L) left and right differential operator. The left and right R-L fractional derivative of order α can be defined, respectively, as follows:

$$
{}_{-\infty}D_t^\alpha \psi\,(t) = \frac{1}{\Gamma\,(m-\alpha)} \frac{d^m}{dt^m} \int_{-\infty}^t (t-\xi)^{m-\alpha-1} \psi\,(\xi)\, d\xi, \quad \text{for} \quad m-1 < \alpha < m, \quad m \in N,
$$

$$(1.19)$$

and

$$
{}_tD_\infty^\alpha \psi\,(t) = \frac{(-1)^m}{\Gamma\,(m-\alpha)} \frac{d^m}{dt^m} \int_t^\infty (\xi-t)^{m-\alpha-1} \psi\,(\xi)\, d\xi, \quad \text{for} \quad m-1 < \alpha < m, \quad m \in N.
$$

$$(1.20)$$

Proof. The proof is very similar to that of Definition 1.6. The only difference is that in Definition 1.6, the integral operator was there, but here the differential operator is present. □

Remark 1.10   It may be noted that the derivative of a constant (c) is not zero in R-L sense, and mathematically it is written as

$$D_t^\alpha c = \frac{ct^{-\alpha}}{\Gamma(1-\alpha)}, \quad 0 < \alpha < 1. \tag{1.21}$$

Proof. If $0 < \alpha < 1$ and $\psi(t) = c$, then Eq. (1.16) reduces to

$$D_t^\alpha c = \frac{1}{\Gamma(1-\alpha)} \frac{d}{dt} \int_0^t (t-\xi)^{1-\alpha-1} c \, d\xi, \quad \text{for} \quad 0 < \alpha < 1. \tag{1.22}$$

Here we have taken $m = 1$ since, from Eq. (1.16), $m - 1 < \alpha < m$. If we compare $m - 1 < \alpha < m$ with $0 < \alpha < 1$, then we obtain $m = 1$. Accordingly, we may get Eq. (1.22). Now, simplifying Eq. (1.22), we have

$$D_t^\alpha c = \frac{c}{\Gamma(1-\alpha)} \frac{d}{dt} \int_0^t (t-\xi)^{-\alpha} d\xi$$

$$= \frac{c}{\Gamma(1-\alpha)} \frac{d}{dt} \left[\frac{-\left((t-\xi)^{-\alpha+1}\right)}{(-\alpha+1)} \right]_0^t = \frac{c}{\Gamma(1-\alpha)} \frac{d}{dt} \left[\frac{t^{-\alpha+1}}{-\alpha+1} \right].$$

$$D_t^\alpha c = \frac{c}{\Gamma(1-\alpha)} (-\alpha+1) \left(\frac{t^{-\alpha}}{-\alpha+1} \right) = \frac{ct^{-\alpha}}{\Gamma(1-\alpha)}.$$

$\square$

Definition 1.11   Caputo fractional differentiation operator.   The fractional-order Caputo sense derivative of an order α of a function $\psi(t)$ is defined as

$$D_t^\alpha \psi(t) = D_t^{\alpha-m} D_t^m \psi(t) = J_t^{m-\alpha} D_t^m \psi(t)$$

$$= \begin{cases} \dfrac{1}{\Gamma(m-\alpha)} \displaystyle\int_0^t (t-\xi)^{m-\alpha-1} \dfrac{d^m \psi(\xi)}{d\xi^m} \, d\xi, & \text{if} \quad m-1 < \alpha < m, \quad m \in N, \\[2mm] \dfrac{d^m}{dt^m} \psi(t), & \text{if} \quad \alpha = m, \quad m \in N. \end{cases}$$

$$\tag{1.23}$$

Proof. Hints: The proof is similar to that of Definition 1.8, but the only difference is that in R-L fractional derivative, the integer-order derivative lies outside the integration, but in the Caputo sense, the integer-order derivative lies inside the integration.

First, Lacroix introduced the integer-order derivatives of a function, which are as follows:

$$D_t^n t^m = \frac{m!}{(m-n)!} t^{m-n} = \frac{\Gamma(m+1)}{\Gamma(m-n+1)} t^{m-n}, \quad \text{for} \quad n \le m. \tag{1.24}$$

Later, he extended this integer-order derivative to fractional-order derivative in the Caputo sense as

$$D_t^\alpha t^\beta = \frac{\Gamma(\beta+1)}{\Gamma(\beta-\alpha+1)} t^{\beta-\alpha}, \quad \text{for} \quad \beta > \alpha - 1, \quad \beta > -1. \tag{1.25}$$

Also, the derivative of a constant in the Caputo sense is zero. Mathematically,

$$D_t^\alpha c = 0.$$

$\square$

Hints: In the Caputo sense derivative, integer-order derivative lies inside the integration, which yields the derivative of a constant as zero that is

$$D_t^\alpha c = \frac{1}{\Gamma(m-\alpha)} \int_0^t (t-\xi)^{m-\alpha-1} \frac{d^m c}{d\xi^m} \, d\xi = \frac{1}{\Gamma(m-\alpha)} \int_0^t (t-\xi)^{m-\alpha-1} 0 \, d\xi = 0. \tag{1.26}$$

Example 1.12

$$\frac{d^{0.5} t}{dt^{0.5}} = \frac{\Gamma(1+1)}{\Gamma(1-0.5+1)} t^{1-0.5} = \frac{1}{\Gamma\left(\frac{3}{2}\right)} t^{0.5} = \frac{2}{\sqrt{\pi}} t^{\frac{1}{2}}.$$

Remark 1.13

(i) The main advantages of Caputo fractional derivative are that the initial conditions for the fractional differential equations are the same form as that of ordinary differential equations. Another advantage is that the Caputo fractional derivative of a constant is zero, while the R-L fractional derivative of a constant is not zero.

(ii) There are several properties in classical derivatives and integrations which follow constant roles. However, these properties may not always hold in the fractional sense. For example, (i) $D_t^\alpha = D_t^{\alpha-m} D_t^m = J_t^{m-\alpha} D_t^m$ and (ii) $D_t^\alpha = D_t^m D_t^{\alpha-m} = D_t^m J_t^{m-\alpha}$ for $m-1 < \alpha < m$ both look equal, but mathematically it is different.

Suppose

$$D_t^{\frac{7}{5}} t = D_t^{\frac{7}{5}-2} D_t^2 t = J_t^{2-\frac{7}{5}} D_t^2 t = J_t^{\frac{3}{5}} D_t^2 t = J_t^{\frac{3}{5}} 0 = 0, \quad \text{(using (i))} \tag{1.27}$$

but

$$D_t^{\frac{7}{5}} t = D_t^2 D_t^{\frac{7}{5}-2} t = D_t^2 J_t^{2-\frac{7}{5}} t = D_t^2 J_t^{\frac{3}{5}} t, \quad \text{(using (ii))}$$

that is

$$D_t^{\frac{7}{5}} t = D_t^2 J_t^{\frac{3}{5}} t. \tag{1.28}$$

Using Eq. (1.14), Eq. (1.28) reduces to

$$D_t^{\frac{7}{5}} t = D_t^2 \frac{\Gamma(1+1)}{\Gamma\left(1+\frac{3}{5}+1\right)} t^{\frac{8}{5}} = \frac{1}{\Gamma(13/5)} D_t^2 t^{\frac{8}{5}} = \frac{24/25}{\Gamma(13/5)} t^{\frac{-2}{5}}. \tag{1.29}$$

On simplifying above Eq. (1.29), we obtain

$$D_t^{\frac{7}{5}} t == \frac{1}{\Gamma(3/5)} t^{\frac{-2}{5}} \neq 0.$$

So, $J_t^{m-\alpha} D_t^m$ is not always equal to $D_t^m J_t^{m-\alpha}$ in a fractional sense.

Note 1.14   In Remark 1.13(ii), $(m-1 =) 1 < \frac{7}{5} < 2 (= m)$, that is, $m = 2$.

Below we incorporate a theorem with respect to the behavior of interchange of derivatives and integration.

Theorem 1.15

(i) $_aD_t^\alpha \left[_aJ_t^\beta \psi(t) \right] = _aD_t^{\alpha-\beta} \psi(t).$

(ii) $_aJ_t^\alpha \left[_aD_t^\beta \psi(t) \right] = _aJ_t^{\alpha-\beta} \psi(t) - \sum_{k=1}^{m} \frac{(t-a)^{\alpha-k}}{\Gamma(\alpha+1-k)} {_aD_t^{\beta-k}} \psi(t)|_{t=a}.$

Here $m = \lceil \beta \rceil + 1$, where $\lceil \beta \rceil$ represents Ceiling function, which means the least integer greater than or equal to β.

Proof.   (i) Hints:

$$_aD_t^\alpha \left[_aJ_t^\beta \psi(t) \right] = \frac{d^n}{dt^n} \left[_aJ_t^{n-\alpha} \left[_aJ_t^\beta \psi(t) \right] \right]$$

$$= \frac{d^n}{dt^n} \left[_aJ_t^{n-(\alpha-\beta)} \right] = _aD_t^{\alpha-\beta} \psi(t).$$

(ii) Hints:

$$I = {_aJ_t^\alpha} \left[_aD_t^\beta \psi(t) \right] = \frac{1}{\Gamma(\alpha)} \int_a^t (t-\xi)^{\alpha-1} \left[_aD_t^\beta \psi(\xi) \right] d\xi.$$

$$I = \frac{1}{\Gamma(\alpha+1)} \int_a^t \frac{d}{dt} (t-\xi)^\alpha \left[_aD_t^\beta \psi(\xi) \right] d\xi.$$

First, Lacroix introduced the integer-order derivatives of a function, which are as follows:

$$D_t^n t^m = \frac{m!}{(m-n)!} t^{m-n} = \frac{\Gamma(m+1)}{\Gamma(m-n+1)} t^{m-n}, \quad \text{for} \quad n \leq m. \tag{1.24}$$

Later, he extended this integer-order derivative to fractional-order derivative in the Caputo sense as

$$D_t^\alpha t^\beta = \frac{\Gamma(\beta+1)}{\Gamma(\beta-\alpha+1)} t^{\beta-\alpha}, \quad \text{for} \quad \beta > \alpha - 1, \quad \beta > -1. \tag{1.25}$$

Also, the derivative of a constant in the Caputo sense is zero. Mathematically,

$$D_t^\alpha c = 0.$$

$\square$

Hints: In the Caputo sense derivative, integer-order derivative lies inside the integration, which yields the derivative of a constant as zero that is

$$D_t^\alpha c = \frac{1}{\Gamma(m-\alpha)} \int_0^t (t-\xi)^{m-\alpha-1} \frac{d^m c}{d\xi^m} \, d\xi = \frac{1}{\Gamma(m-\alpha)} \int_0^t (t-\xi)^{m-\alpha-1} 0 \, d\xi = 0. \tag{1.26}$$

Example 1.12

$$\frac{d^{0.5} t}{dt^{0.5}} = \frac{\Gamma(1+1)}{\Gamma(1-0.5+1)} t^{1-0.5} = \frac{1}{\Gamma\left(\frac{3}{2}\right)} t^{0.5} = \frac{2}{\sqrt{\pi}} t^{\frac{1}{2}}.$$

Remark 1.13

(i) The main advantages of Caputo fractional derivative are that the initial conditions for the fractional differential equations are the same form as that of ordinary differential equations. Another advantage is that the Caputo fractional derivative of a constant is zero, while the R-L fractional derivative of a constant is not zero.

(ii) There are several properties in classical derivatives and integrations which follow constant roles. However, these properties may not always hold in the fractional sense. For example, (i) $D_t^\alpha = D_t^{\alpha-m} D_t^m = J_t^{m-\alpha} D_t^m$ and (ii) $D_t^\alpha = D_t^m D_t^{\alpha-m} = D_t^m J_t^{m-\alpha}$ for $m-1 < \alpha < m$ both look equal, but mathematically it is different.

Suppose

$$D_t^{\frac{7}{5}} t = D_t^{\frac{7}{5}-2} D_t^2 t = J_t^{2-\frac{7}{5}} D_t^2 t = J_t^{\frac{3}{5}} D_t^2 t = J_t^{\frac{3}{5}} 0 = 0, \quad \text{(using (i))} \tag{1.27}$$

but

$$D_t^{\frac{7}{5}} t = D_t^2 D_t^{\frac{7}{5}-2} t = D_t^2 J_t^{2-\frac{7}{5}} t = D_t^2 J_t^{\frac{3}{5}} t, \quad \text{(using (ii))}$$

that is

$$D_t^{\frac{7}{5}} t = D_t^2 J_t^{\frac{3}{5}} t. \tag{1.28}$$

Using Eq. (1.14), Eq. (1.28) reduces to

$$D_t^{\frac{7}{5}} t = D_t^2 \frac{\Gamma(1+1)}{\Gamma\left(1+\frac{3}{5}+1\right)} t^{\frac{8}{5}} = \frac{1}{\Gamma(13/5)} D_t^2 t^{\frac{8}{5}} = \frac{24/25}{\Gamma(13/5)} t^{\frac{-2}{5}}. \tag{1.29}$$

On simplifying above Eq. (1.29), we obtain

$$D_t^{\frac{7}{5}} t == \frac{1}{\Gamma(3/5)} t^{\frac{-2}{5}} \neq 0.$$

So, $J_t^{m-\alpha} D_t^m$ is not always equal to $D_t^m J_t^{m-\alpha}$ in a fractional sense.

Note 1.14 In Remark 1.13(ii), $(m-1=)1 < \frac{7}{5} < 2\,(=m)$, that is, $m=2$.

Below we incorporate a theorem with respect to the behavior of interchange of derivatives and integration.

Theorem 1.15

(i) $_aD_t^\alpha \left[_aJ_t^\beta \psi(t) \right] = {_aD_t^{\alpha-\beta}} \psi(t).$

(ii) $_aJ_t^\alpha \left[_aD_t^\beta \psi(t) \right] = {_aJ_t^{\alpha-\beta}} \psi(t) - \sum_{k=1}^m \frac{(t-a)^{\alpha-k}}{\Gamma(\alpha+1-k)} {_aD_t^{\beta-k}} \psi(t)|_{t=a}.$

Here $m = \lceil \beta \rceil + 1$, *where* $\lceil \beta \rceil$ *represents Ceiling function, which means the least integer greater than or equal to* β.

Proof. (i) Hints:

$$_aD_t^\alpha \left[_aJ_t^\beta \psi(t) \right] = \frac{d^n}{dt^n} \left[_aJ_t^{n-\alpha} \left[_aJ_t^\beta \psi(t) \right] \right]$$
$$= \frac{d^n}{dt^n} \left[_aJ_t^{n-(\alpha-\beta)} \right] = {_aD_t^{\alpha-\beta}} \psi(t).$$

(ii) Hints:

$$I = {_aJ_t^\alpha} \left[_aD_t^\beta \psi(t) \right] = \frac{1}{\Gamma(\alpha)} \int_a^t (t-\xi)^{\alpha-1} \left[_aD_t^\beta \psi(\xi) \right] d\xi.$$
$$I = \frac{1}{\Gamma(\alpha+1)} \int_a^t \frac{d}{dt} (t-\xi)^\alpha \left[_aD_t^\beta \psi(\xi) \right] d\xi.$$

Applying integration by parts, we have

$$I = -\frac{(t-a)^\alpha}{\Gamma(\alpha+1)} \left[{}_a D_t^\beta \psi(\xi) \Big|_{t=a} \right] + \frac{1}{\Gamma(\alpha+2)} \int_a^t \frac{d}{dt} (t-\xi)^{\alpha+1} \left[{}_a D_t^{\beta-1} \psi(\xi) \right] d\xi.$$

Again, successively applying integration by part, we may get the desired result.

$\square$

Definition 1.16   Grunwald–Letnikov differential operator.   The differential operator D_t^α of order α in the Grunwald–Letnikov sense is defined as

$$_a D_t^\alpha \psi(t) = \lim_{h \to 0} \frac{1}{h^\alpha} \sum_{r=0}^{\left[\frac{t-a}{h}\right]} (-1)^r \frac{\Gamma(\alpha+1)}{r! \Gamma(\alpha-r+1)} \psi(x-rh). \tag{1.30}$$

Proof.   Let $\psi(x) \in [a,b]$. Then the first-order derivative of the function $\psi(x)$ is defined as

$$\psi^{(1)}(t) = \frac{d\psi}{dt} = \lim_{h \to 0} \frac{\psi(t) - \psi(t-h)}{h}.$$

Again applying a derivative operator on the above equation, we obtain

$$\psi^2(t) = \frac{d^2\psi}{dt^2} = \lim_{h \to 0} \frac{\psi(t) - 2\psi(t-h) + \psi(t-2h)}{h^2}.$$

With the help of the method of induction, we may write nth-order derivative as

$$\psi^n(t) = \frac{d^n\psi}{dt^n} = \lim_{h \to 0} \frac{1}{h^n} \sum_{r=0}^{n} (-1)^r \binom{n}{r} \psi(t-rh), \tag{1.31}$$

where $a \le t \le b, h = \frac{t-a}{n}$, and $\binom{n}{r} = \frac{n(n-1)(n-2)\cdot(n-r+1)}{r!}$.

Now, we extend the nth order derivative to fractional-order derivate as follows:

$$\psi^\alpha(t) = \frac{d^\alpha\psi}{dt^\alpha} = \lim_{h \to 0} \frac{1}{h^\alpha} \sum_{r=0}^{\alpha} (-1)^r \binom{\alpha}{r} \psi(t-rh), \tag{1.32}$$

where $\binom{\alpha}{r} = \frac{\alpha!}{r!(\alpha-r)!} = \frac{\Gamma(1+\alpha)}{r!\Gamma(\alpha-r+1)}$ and $h = \frac{t-a}{n} \Rightarrow n = \frac{t-a}{h}$. Since n is replaced by α and α is a non-integer, we can write $\alpha = \left[\frac{t-a}{h}\right]$. $\square$

Definition 1.17   Grunwald–Letnikov integral operator.   The fractional Grunwald–Letnikov integral operator is defined as

$$_a D_t^{-\alpha} \psi(t) = \lim_{h \to 0} h^\alpha \sum_{r=0}^{[\frac{t-a}{h}]} \frac{\Gamma(\alpha + r)}{r! \Gamma(\alpha)} \psi(x - rh). \tag{1.33}$$

Proof. Hints: Replacing α by $-\alpha$ of the Eq. (1.30), we get

$$_a D_t^{-\alpha} \psi(t) = \lim_{h \to 0} h^\alpha \sum_{r=0}^{[\frac{t-a}{h}]} (-1)^r \frac{\Gamma(-\alpha + 1)}{r! \Gamma(-\alpha - r + 1)} \psi(x - rh); \tag{1.34}$$

we know that

$$\frac{\Gamma(-\alpha + 1)}{r! \Gamma(-\alpha - r + 1)} = \binom{-\alpha}{r} = \frac{-\alpha(-\alpha - 1)(-\alpha - 2) \ldots (-\alpha - r + 1)}{r!}$$

$$= \frac{(-1)^r \alpha(\alpha + 1)(\alpha + 2) \ldots (\alpha + r - 1)(\alpha - 1)!}{r!(\alpha - 1)!} = \frac{(-1)^r \Gamma(\alpha + r)}{r! \Gamma(\alpha)}. \tag{1.35}$$

Substituting Eq. (1.35) into Eq. (1.34), we get Eq. (1.33).   □

Definition 1.18   Riesz fractional integral operator.   The Riesz fractional integration of a function $\psi(t)$ of order α is defined as

$$_0^R J_t^\alpha \psi(t) = c_\alpha \left(_{-\infty} J_t^\alpha + {}_t J_\infty^\alpha \right) \psi(t) = \frac{c_\alpha}{\Gamma(\alpha)} \int_{-\infty}^{\infty} |t - \xi|^{\alpha - 1} \psi(\xi) \, d\xi, \tag{1.36}$$

where $c_\alpha = \frac{1}{2 \cos(\frac{\alpha \pi}{2})}$, $\alpha \neq 1$, and $m - 1 < \alpha \leq m$.

Definition 1.19   Riesz fractional differential operator.   The Riesz fractional differentiation of a function $\psi(t)$ of order α on the infinite domain $-\infty < t < \infty$ is defined as

$$\frac{d^\alpha \psi(t)}{d|t|^\alpha} = -c_\alpha \left(_{-\infty} D_t^\alpha + {}_t D_\infty^\alpha \right) \psi(t), \tag{1.37}$$

where $c_\alpha = \frac{1}{2 \cos(\frac{\alpha \pi}{2})}$, $\alpha \neq 1$ and $m - 1 < \alpha \leq m$.

Here, various definitions of fractional order derivatives are given. But, in the subsequent chapters, the practical problems of time-fractional order in Caputo sense are solved in an uncertain environment.

1.4 REFERENCES

[1] A. A. Kilbas, H. M. Srivastava, and J. J. Trujillo. *Theory and Application of Fractional Differential Equations*. Elsevier Science B.V., Amsterdam, 2006.

[2] V. S. Kiryakova. *Generalized Fractional Calculus and Applications*. Longman Scientific and Technical, England, 1993.

[3] V. Lakshmikantham and R. N. Mohapatra. *Theory of Fuzzy Differential Equations and Applications*. Taylor & Francis, London, 2003. DOI: 10.1201/9780203011386.

[4] K. S. Miller and B. Ross. *An Introduction to the Fractional Calculus and Fractional Differential Equations*. John Wiley & Sons, NY, 1993.

[5] K. B. Oldham and J. Spanier. *The Fractional Calculus*. Academic Press, New York, 1974.

[6] I. Podlubny. *Fractional Differential Equations*. Academic Press, New York, 1999. 4

[7] S. Chakraverty, S. Tapaswini, and D. Behera. *Fuzzy Arbitrary Order System: Fuzzy Fractional Differential Equations and Applications*. John Wiley & Sons, NJ, 2016. DOI: 10.1002/9781119004233.

[8] S. Das. *Functional Fractional Calculus*. Springer Science & Business Media, 2011. DOI: 10.1007/978-3-642-20545-3.

[9] D. Baleanu, J. A. T. Machado, and A. C. J. Luo. *Fractional Dynamics, and Control*. Springer, 2012. DOI: 10.1007/978-1-4614-0457-6.

[10] D. Baleanu, K. Diethelm, E. Scalas, and J. J. Trujillo. *Fractional Calculus: Models and Numerical Methods*. World Scientific Publishing Company, 2012. DOI: 10.1142/10044.

[11] A. Atangana. *Derivative with a New Parameter Theory, Methods and Applications*. Elsevier, London Wall, London, 2016.

[12] R. M. Jena, S. Chakraverty, and D. Baleanu. On the solution of imprecisely defined nonlinear time-fractional dynamical model of marriage. *Mathematics*, 7:689–704, 2019. DOI: 10.3390/math7080689.

[13] R. M. Jena, S. Chakraverty, and D. Baleanu. On new solutions of time-fractional wave equations arising in Shallow water wave propagation. *Mathematics*, 7:722–734, 2019.

[14] R. M. Jena and S. Chakraverty. Solving time-fractional Navier–Stokes equations using homotopy perturbation Elzaki transform. *SN Applied Sciences*, 1(1):16, 2019. DOI: 10.1007/s42452-018-0016-9.

12 REFERENCES

[15] R. M. Jena and S. Chakraverty. Residual power series method for solving time-fractional model of vibration equation of large membranes. *Journal of Applied and Computational Mechanics*, 5:603–615, 2019. DOI: 10.22055/jacm.2018.26668.1347.

[16] R. M. Jena and S. Chakraverty. A new iterative method based solution for fractional Black–Scholes option pricing equations (BSOPE). *SN Applied Sciences*, 1:95–105, 2019. DOI: 10.1007/s42452-018-0106-8.

[17] R. M. Jena and S. Chakraverty. Analytical solution of Bagley–Torvik equations using Sumudu transformation method. *SN Applied Sciences*, 1(3):246, 2019. DOI: 10.1007/s42452-019-0259-0.

[18] R. M. Jena, S. Chakraverty, and S. K. Jena, Dynamic response analysis of fractionally damped beams subjected to external loads using homotopy analysis method. *Journal of Applied and Computational Mechanics*, 5:355–366, 2019. DOI: 10.22055/jacm.2019.27592.1419.

CHAPTER 2

Preliminaries of Fuzzy Set Theory

2.1 INTRODUCTION

In this chapter, we present the definitions of fuzzy set, fuzzy numbers (viz. triangular, trapezoidal, Gaussian fuzzy numbers), double parametric form, type of differentiability, and fuzzy/interval arithmetic. Further, theorems/lemma related to fuzzy fractional differential equations which are significant to the subsequent chapters are included. Excellent books related to various aspects of fuzzy set theory are Zimmermann [1], Jaulin et al. [2], Ross [3], Hanss [4], Moore [5], Chakraverty [6], and Chakravert et al. [7–9], which may be referred to for understating further details.

Before defining the fuzzy set, let us first introduce the concepts of interval.

2.2 INTERVAL

An interval $\tilde{\psi}$ is defined in Eq. (2.1) by $[\underline{\psi}, \overline{\psi}]$ on the set of real number $\Re$ is given by

$$\tilde{\psi} = [\underline{\psi}, \overline{\psi}] = \left\{ \psi \in \Re : \underline{\psi} \leq \psi \leq \overline{\psi} \right\}, \tag{2.1}$$

where $\underline{\psi}$ and $\overline{\psi}$ represent the left and right endpoints of the interval.

Although there are various types of the interval, such as open and half-open intervals, we have considered closed intervals throughout the book.

The two arbitrary intervals $\tilde{\psi} = [\underline{\psi}, \overline{\psi}]$ and $\tilde{\xi} = [\underline{\xi}, \overline{\xi}]$ are said to be equal if both the intervals are in the same set. Mathematically one may write

$$\tilde{\psi} = \tilde{\xi} \quad \text{if and only if} \quad \underline{\psi} = \underline{\xi} \quad \text{and} \quad \overline{\psi} = \overline{\xi}. \tag{2.2}$$

2.3 INTERVAL ARITHMETIC

For the given two arbitrary intervals $\tilde{\psi} = [\underline{\psi}, \overline{\psi}]$ and $\tilde{\xi} = [\underline{\xi}, \overline{\xi}]$, the interval arithmetic operations addition (+), subtraction (−), multiplication (×), and division (/) are defined as follows:

(i) $\tilde{\psi} + \tilde{\xi} = [\underline{\psi} + \underline{\xi}, \overline{\psi} + \overline{\xi}]$.

(ii) $\tilde{\psi} - \tilde{\xi} = [\underline{\psi} - \overline{\xi}, \overline{\psi} - \underline{\xi}]$.

(iii) $\tilde{\psi} \times \tilde{\xi} = \left[\min \left(\underline{\psi} \times \underline{\xi}, \ \underline{\psi} \times \overline{\xi}, \overline{\psi} \times \underline{\xi}, \ \overline{\psi} \times \overline{\xi} \right), \ \max \left(\underline{\psi} \times \underline{\xi}, \ \underline{\psi} \times \overline{\xi}, \overline{\psi} \times \underline{\xi}, \ \overline{\psi} \times \overline{\xi} \right) \right].$

(iv) $\frac{\tilde{\psi}}{\tilde{\xi}} = [\underline{\psi}, \overline{\psi}] \times \left[\frac{1}{\overline{\xi}}, \frac{1}{\underline{\xi}} \right].$ if $0 \notin \tilde{\xi}.$

(v) $k\tilde{\psi} = \begin{cases} \left[k\overline{\psi}, k\underline{\psi} \right], & k < 0, \\ \left[k\underline{\psi}, k\overline{\psi} \right], & k \geq 0, \end{cases}$ where k is a real number.

2.4   FUZZY SET

A fuzzy set $\tilde{F}$ is a set of ordered pairs consisting of the elements x of a universal set X and a certain degree of pre-assumed membership values $\mu(x)$ of the form

$$\tilde{F} = \{(x, \ \mu(x)) : \ x \in X, \ \mu(x) \in [0, \ 1]\},$$

where $\mu(x)$ is the membership function of the fuzzy set and is piecewise continuous.

2.5 γ-CUT

The γ-cut associated with a fuzzy number $\tilde{\psi}$ is the crisp set

$$\psi_\gamma = \left\{ x \in \Re : \ \mu_{\tilde{\psi}}(x) \geq \gamma \right\}.$$

2.6   FUZZY NUMBER

A fuzzy number $\tilde{\psi}$ is a convex normalized fuzzy set $\tilde{\psi}$ of the real line $\Re$ such that

$$\left\{ \mu_{\tilde{\psi}}(x) : \Re \to [0, \ 1], \ \forall \, x \in \Re \right\},$$

where $\mu_{\tilde{\psi}}$ is a membership function and is piecewise continuous.

There exist various types of fuzzy numbers. However, only three types of fuzzy number, viz. triangular, trapezoidal, and Gaussian fuzzy numbers, are defined here.

Definition 2.1   Triangular Fuzzy Number (TFN).   A TFN $\tilde{\psi}$ is a convex normalized fuzzy set $\tilde{\psi}$ of the real line $\Re$ such that

(a) There exists exactly one $x_0 \in \Re$ with $\mu_{\tilde{\psi}}(x_0) = 1$ (x_0 is called the mean value of $\tilde{\psi}$), where $\mu_{\tilde{\psi}}$ is called the membership function of the fuzzy set.

(b) $\mu_{\tilde{\psi}}(x)$ is piecewise continuous.

Let us consider the TFN, $\tilde{\psi} = (a_1, \ b_1, \ c_1)$. Then, the membership function $\mu_{\tilde{\psi}}$ of a TFN $\tilde{\psi} = (a_1, \ b_1, \ c_1)$ is defined as follows:

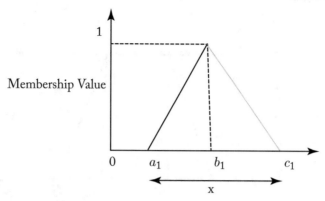

Figure 2.1: Triangular fuzzy number.

$$\mu_{\tilde{\psi}}(x) = \begin{cases} 0, & x \leq a_1, \\ \dfrac{x - a_1}{b_1 - a_1}, & a_1 \leq x \leq b_1, \\ \dfrac{c_1 - x}{c_1 - b_1}, & b_1 \leq x \leq c_1, \\ 0, & x \geq c_1. \end{cases}$$

The TFN $\tilde{\psi} = (a_1, b_1, c_1)$ may be represented by an ordered pair of functions through γ-cut approach viz. $[\underline{\psi}(\gamma), \overline{\psi}(\gamma)] = [(b_1 - a_1)\gamma + a_1, -(c_1 - b_1)\gamma + c_1]$, where $\gamma \in [0, 1]$ (see Fig. 2.1).

Definition 2.2   Trapezoidal Fuzzy Number (TrFN).   We consider the arbitrary TrFN $\tilde{\psi} = (a_1, b_1, c_1, d_1)$. Then the membership function $\mu_{\tilde{\psi}}$ of a TrFN $\tilde{\psi} = (a_1, b_1, c_1, d_1)$ is defined as follows:

$$\mu_{\tilde{\psi}}(x) = \begin{cases} 0, & x \leq a_1, \\ \dfrac{x - a_1}{b_1 - a_1}, & a_1 \leq x \leq b_1, \\ 1 & b_1 \leq x \leq c_1, \\ \dfrac{d_1 - x}{d_1 - c_1}, & c_1 \leq x \leq d_1, \\ 0, & x \geq d_1. \end{cases}$$

The TrFN $\tilde{\psi} = (a_1, b_1, c_1, d_1)$ may be represented by an ordered pair of functions through γ-cut approach viz. $[\underline{\psi}(\gamma), \overline{\psi}(\gamma)] = [(b_1 - a_1)\gamma + a_1, -(d_1 - c_1)\gamma + d_1]$, where $\gamma \in [0, 1]$ (see Fig. 2.2).

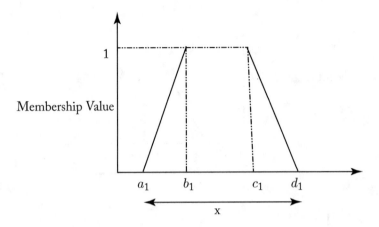

Figure 2.2: Trapezoidal fuzzy number.

Definition 2.3   Gaussian Fuzzy Number (GFN).   Let us consider an arbitrary asymmetrical GFN, $\tilde{\psi} = (\eta,\ \lambda_1,\ \lambda_2)$. The corresponding membership function may be written as

$$\mu_{\tilde{\psi}}(x) = \begin{cases} \exp\left(\frac{-(x-\eta)^2}{2\lambda_1^2}\right), & \text{for}\quad x \leq \eta \\[2mm] \exp\left(\frac{-(x-\eta)^2}{2\lambda_2^2}\right), & \text{for}\quad x \geq \eta \end{cases} \quad \text{for}\quad \forall x \in \Re, \qquad (2.3)$$

where η denotes the modal value and λ_1, λ_2 denote the left-hand and right-hand fuzziness, respectively. For symmetric GFN, both left-hand and right-hand fuzziness are equal, that is $\lambda_1 = \lambda_2 = \lambda$. So, the symmetric GFN is written as $\tilde{\psi} = (\eta,\ \lambda, \lambda)$ and the corresponding membership function may be defined as $\mu_{\tilde{\psi}}(x) = \exp\left(\frac{-(x-\eta)^2}{2\lambda^2}\right) = \exp\left\{-\beta\,(x-\eta)^2\right\},\ \forall x \in \Re$ where $\beta = \frac{1}{2\lambda^2}$. The symmetric GFN may be expressed in parametric and can be represented as

$$\left[\underline{\psi}(\gamma),\ \overline{\psi}(\gamma)\right] = \left[\eta - \sqrt{-\frac{\log_e \gamma}{\beta}},\ \eta + \sqrt{-\frac{\log_e \gamma}{\beta}}\right], \quad \text{where}\quad \gamma \in [0,\ 1].$$

For all the aforementioned fuzzy numbers, the lower and upper bounds of the fuzzy numbers satisfy the below statements:

(i) $\underline{\psi}(\gamma)$ is a bounded left-continuous nondecreasing function over $[0,\ 1]$.

(ii) $\overline{\psi}(\gamma)$ is a bounded right-continuous nonincreasing function over $[0,\ 1]$.

(iii) $\underline{\psi}(\gamma) \leq \overline{\psi}(\gamma)$, where $0 \leq \gamma \leq 1$.

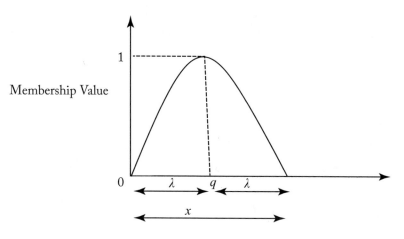

Figure 2.3: Gaussian fuzzy number.

2.7   DOUBLE PARAMETRIC FORM (DPF) OF FUZZY NUMBER

Implementing γ-cut approach as discussed in Definitions 2.3–2.6 for all the fuzzy numbers, we have $\tilde{\psi} = [\underline{\psi}(\gamma), \overline{\psi}(\gamma)]$. Now, we may write the above form by using another parameter β as

$$\tilde{\psi}(\gamma, \beta) = \beta\left(\overline{\psi}(\gamma) - \underline{\psi}(\gamma)\right) + \underline{\psi}(\gamma), \quad \text{where} \quad \beta \text{ and } \gamma \in [0, 1],$$

which is named DPF as it involves two parameters viz. γ and β.

In order to obtain lower and upper bounds solution in single parametric form, we may substitute $\beta = 0$ and $\beta = 1$, respectively. Mathematically, these can be represented as $\tilde{\psi}(\gamma, 0) = \underline{\psi}(\gamma)$ and $\tilde{\psi}(\gamma, 1) = \overline{\psi}(\gamma)$.

Definition 2.4 Fuzzy Center.   Fuzzy center of an arbitrary fuzzy number $\tilde{\psi} = [\underline{\psi}(\gamma), \overline{\psi}(\gamma)]$ is defined as $\tilde{\psi}^C = \frac{\underline{\psi}(\gamma) + \overline{\psi}(\gamma)}{2}, \forall\ 0 \leq \gamma \leq 1$.

Definition 2.5 Fuzzy Radius.   Fuzzy radius of an arbitrary fuzzy number $\tilde{\psi} = [\underline{\psi}(\gamma), \overline{\psi}(\gamma)]$ is defined as $\tilde{\psi}^R = \frac{\overline{\psi}(\gamma) - \underline{\psi}(\gamma)}{2}, \forall\ 0 \leq \gamma \leq 1$.

Definition 2.6 Fuzzy Width.   Fuzzy width of an arbitrary fuzzy number $\tilde{\psi} = [\underline{\psi}(\gamma), \overline{\psi}(\gamma)]$ is defined as $\left|\overline{\psi}(\gamma) - \underline{\psi}(\gamma)\right|, \forall\ 0 \leq \gamma \leq 1$.

Definition 2.7   Fuzzy Arithmetic. For the given two arbitrary fuzzy numbers $\tilde{\psi} = [\underline{\psi}(\gamma), \overline{\psi}(\gamma)]$ and $\tilde{\xi} = [\underline{\xi}(\gamma), \overline{\xi}(\gamma)]$, the fuzzy arithmetic is similar to the interval arithmetic, which is defined as follows:

(i) $\tilde{\psi} + \tilde{\xi} = [\underline{\psi}(\gamma) + \underline{\xi}(\gamma), \overline{\psi}(\gamma) + \overline{\xi}(\gamma)]$.

(ii) $\tilde{\psi} - \tilde{\xi} = [\underline{\psi}(\gamma) - \overline{\xi}(\gamma), \overline{\psi}(\gamma) - \underline{\xi}(\gamma)]$.

(iii) $\tilde{\psi} \times \tilde{\xi} = \begin{bmatrix} \min\left(\underline{\psi}(\gamma) \times \underline{\xi}(\gamma), \ \underline{\psi}(\gamma) \times \overline{\xi}(\gamma), \ \overline{\psi}(\gamma) \times \underline{\xi}(\gamma), \ \overline{\psi}(\gamma) \times \overline{\xi}(\gamma)\right), \\ \max\left(\underline{\psi}(\gamma) \times \underline{\xi}(\gamma), \ \underline{\psi}(\gamma) \times \overline{\xi}(\gamma), \ \overline{\psi}(\gamma) \times \underline{\xi}(\gamma), \ \overline{\psi}(\gamma) \times \overline{\xi}(\gamma)\right) \end{bmatrix}$.

(iv) $\frac{\tilde{\psi}}{\tilde{\xi}} = [\underline{\psi}(\gamma), \overline{\psi}(\gamma)] \times \left[\frac{1}{\overline{\xi}(\gamma)}, \frac{1}{\underline{\xi}(\gamma)}\right]$ if $0 \notin \tilde{\xi}(\gamma)$.

(v) $k\tilde{\psi} = \begin{cases} [k\overline{\psi}(\gamma), k\underline{\psi}(\gamma)], & k < 0, \\ [k\underline{\psi}(\gamma), k\overline{\psi}(\gamma)], & k \geq 0. \end{cases}$ where k is a real number.

Definition 2.8   Let $\psi : (a, b) \to R_\psi$ and $t_0 = (a, b)$. Then ψ is called differential at t_0 if there exist $\psi'(t_0) \in R_\psi$ such that

(i) for all $h > 0$ sufficiently close to zero, the Hukuhara difference $\psi(t_0 + h) \ominus \psi(t_0)$ and $\psi(t_0) \ominus \psi(t_0 - h)$ exist and

$$\lim_{h \to 0^+} \frac{\psi(t_0 + h) \ominus \psi(t_0)}{h} = \lim_{h \to 0^+} \frac{\psi(t_0) \ominus \psi(t_0 - h)}{h} = \psi'(t_0), \qquad (2.4)$$

or

(ii) for all $h > 0$ sufficiently close to zero, the Hukuhara difference $\psi(t_0) \ominus \psi(t_0 + h)$ and $\psi(t_0 - h) \ominus \psi(t_0)$ exist and

$$\lim_{h \to 0^+} \frac{\psi(t_0) \ominus \psi(t_0 + h)}{-h} = \lim_{h \to 0^+} \frac{\psi(t_0 - h) \ominus \psi(t_0)}{-h} = \psi'(t_0). \qquad (2.5)$$

Interested authors may see the references Chalco-Cano and Roman-Flores [10], Khastan et al. [11] for a better understanding of the fuzzy differentiation and Hukuhara difference.

Chalco-Cano and Roman-Flores [10] used the above definition to obtain the following results, which are given in terms of a theorem.

Theorem 2.9   *Let $\psi : (a, b) \to R_\psi$ and denote $\psi(t; \gamma) = [\underline{\psi}(t; \gamma), \overline{\psi}(t; \gamma)]$ for each $\gamma \in [0, 1]$.*

(i) *If ψ is differential of the first type (i) (see Definition 2.8(i)), then $\underline{\psi}(t; \gamma)$ and $\overline{\psi}(t; \gamma)$ are differentiable functions, and we have $\psi'(t; \gamma) = [\underline{\psi}'(t; \gamma), \overline{\psi}'(t; \gamma)]$.*

(ii) *If ψ is differential of the second type (ii) (see Definition 2.8(ii)), then $\underline{\psi}(t;\gamma)$ and $\overline{\psi}(t;\gamma)$ are differentiable functions, and we have $\psi'(t;\gamma) = [\overline{\psi}'(t;\gamma), \underline{\psi}'(t;\gamma)]$.*

Proof. One may see the proof of the theorem in reference [10].   □

2.8   FUZZY FRACTIONAL REIMANN–LIOUVILLE INTEGRAL

The fuzzy fractional Reimann–Liouville operator of a fuzzy valued function $\tilde{\psi}(t;\gamma)$ of order α, based on its γ-cut representation, is defined as follows:

$$J_t^\alpha \tilde{\psi}(t;\gamma) = \left[J_t^\alpha \underline{\psi}(t;\gamma), J_t^\alpha \overline{\psi}(t;\gamma) \right], t > 0,$$

where,

$$\begin{cases} J_t^\alpha \underline{\psi}(t;\gamma) = \dfrac{1}{\Gamma(\alpha)} \displaystyle\int_0^t (t-\xi)^{\alpha-1} \underline{\psi}(\xi;\gamma)\, d\xi, \text{ and} \\ J_t^\alpha \overline{\psi}(t;\gamma) = \dfrac{1}{\Gamma(\alpha)} \displaystyle\int_0^t (t-\xi)^{\alpha-1} \overline{\psi}(\xi;\gamma)\, d\xi, \end{cases} \qquad t > 0. \qquad (2.6)$$

2.9   FUZZY FRACTIONAL CAPUTO DERIVATIVE

Let $\tilde{\psi}(t;\gamma)$ be a fuzzy valued function and $\tilde{\psi}(t;\gamma) = \left[\underline{\psi}(t;\gamma), \overline{\psi}(t;\gamma) \right]$, for $\gamma \in [0,1]$, $0 < \alpha < 1$, and $t \in (a,b)$.

(i) If $\tilde{\psi}(t;\gamma)$ is a fuzzy fractional differentiable function defined in Caputo sense in the first form, then

$$D_t^\alpha \tilde{\psi}(t;\gamma) = \left[D_t^\alpha \underline{\psi}(t;\gamma), D_t^\alpha \overline{\psi}(t;\gamma) \right]. \qquad (2.7)$$

(ii) If $\tilde{\psi}(t;\gamma)$ is a fuzzy fractional differentiable function defined in Caputo sense in the second form, then

$$D_t^\alpha \tilde{\psi}(t;\gamma) = \left[D_t^\alpha \overline{\psi}(t;\gamma), D_t^\alpha \underline{\psi}(t;\gamma) \right], \qquad (2.8)$$

where,

$$D_t^\alpha \underline{\psi}(t;\gamma) = \frac{1}{\Gamma(m-\alpha)} \int_0^t (t-\xi)^{m-\alpha-1} \frac{d^m \underline{\psi}(\xi;\gamma)}{d\xi^m}\, d\xi, \quad m-1 < \alpha < m, m \in N,$$

$$D_t^\alpha \overline{\psi}(t;\gamma) = \frac{1}{\Gamma(m-\alpha)} \int_0^t (t-\xi)^{m-\alpha-1} \frac{d^m \overline{\psi}(\xi;\gamma)}{d\xi^m}\, d\xi, \quad m-1 < \alpha < m, m \in N,$$

$$\frac{d^m}{dt^m} \tilde{\psi}(t;\gamma), \quad \alpha = m, m \in N.$$

2.10   FRACTIONAL INITIAL VALUE PROBLEM (FIVP)

Let us consider the following FIVP defined in the Caputo sense:

$$D_t^\alpha \psi(t) = f(t, \psi),$$

(2.9)

subject to initial condition:

$$\psi(0) = \psi_0, \ t \in [a, b], \ \alpha \in (0, 1).$$

Next, we combine the FIVP and with uncertainty to obtain a new type of dynamical system called fuzzy fractional-order initial value problem.

2.11   FUZZY FRACTIONAL INITIAL VALUE PROBLEM (FFIVP)

Let us consider the following FFIVP defined in the Caputo sense:

$$D_t^\alpha \tilde{\psi}(t) = f(t, \tilde{\psi}),$$

(2.10)

subject to fuzzy initial condition:

$$\tilde{\psi}(0) = \tilde{\psi}_0, \ t \in [a, b], \ \alpha \in (0, 1).$$

Now, Eq. (2.10) can be considered equivalent to the following initial value problem:

$$\left[D_t^\alpha \underline{\psi}(t), \ D_t^\alpha \overline{\psi}(t) \right] = \left[\underline{f}(t, \underline{\psi}), \ \overline{f}(t, \overline{\psi}) \right],$$

subject to the fuzzy initial condition:

$$\left[\underline{\psi}(0), \overline{\psi}(0) \right] = \left[\underline{\psi}_0, \overline{\psi}_0 \right].$$

2.12   FRACTIONAL BOUNDARY VALUE PROBLEM (FBVP)

Let us consider the following FBVP defined in the Caputo sense:

$$D_t^\alpha \psi(t) = f(t, \psi),$$

subject to boundary condition:

$$\psi(a) = \psi_a, \ \psi(b) = \psi_b, \ \alpha \in (0, 1).$$

2.13   FUZZY FRACTIONAL BOUNDARY VALUE PROBLEM (FFBVP)

Let us consider the following FFBVP defined in the Caputo sense:

$$D_t^\alpha \tilde{\psi}(t) = f\left(t, \tilde{\psi}\right), \tag{2.11}$$

subject to fuzzy boundary condition:

$$\tilde{\psi}(a) = \tilde{\psi}_a, \quad \tilde{\psi}(b) = \tilde{\psi}_b.$$

Now, Eq. (2.11) can be considered equivalent to the following boundary value problem:

$$\left[D_t^\alpha \underline{\psi}(t),\, D_t^\alpha \overline{\psi}(t)\right]\left[\underline{f}\left(t, \underline{\psi}\right),\, \overline{f}\left(t, \overline{\psi}\right)\right],$$

subject to the fuzzy boundary condition:

$$\left[\underline{\psi}(a),\, \overline{\psi}(a)\right] = \left[\underline{\psi}_a,\, \overline{\psi_a}\right] \quad \text{and} \quad \left[\underline{\psi}(b),\, \overline{\psi}(b)\right] = \left[\underline{\psi}_b,\, \overline{\psi_b}\right].$$

Lemma 2.10   *If $\tilde{u}(t) = (x(t),\, y(t),\, z(t))$ is a TFN valued function and $\tilde{u}$ is Hukuhara differentiable [13], then $\tilde{u}' = (x',\, y',\, z')$.*

By using Hukuhara differentiable, our main aim is to solve the following fuzzy initial value problem:

$$\tilde{x}'(t) = f(t, \tilde{x}), \tag{2.12}$$

subject to the triangular fuzzy initial condition:

$$\tilde{x}(t_0) = \tilde{x}_0,$$

where, $\tilde{x}_0 = \left(\underline{x_0}, x_0, \overline{x_0}\right) \in R$, $\tilde{x}(t) = (\underline{u}, u, \overline{u}) \in R$, and

$$f : [t_0, t_0 + a] \times R \to R,\ f(t, (\underline{u}, u, \overline{u})) = \left(\underline{f}(t, \underline{u}, u, \overline{u}),\, f(t, \underline{u}, u, \overline{u}),\, \overline{f}(t, \underline{u}, u, \overline{u})\right).$$

We can transfer this into the following system of ordinary differential equations:

$$\begin{cases} \underline{u} = \underline{f}(t, \underline{u}, u, \overline{u}) \\ u = f(t, \underline{u}, u, \overline{u}) \\ \overline{u} = \overline{f}(t, \underline{u}, u, \overline{u}) \\ \underline{u}(0) = \underline{x_0},\ u(0) = x_0,\ \overline{u}(0) = \overline{x_0}. \end{cases} \tag{2.13}$$

2.14 REFERENCES

[1] H. J. Zimmermann. *Fuzzy Set Theory and its Application*. Kluwer Academic Publishers, Boston/Dordrecht/London, 2001. DOI: 10.1007/978-94-015-7949-0. 13

[2] L. Jaulin, M. Kieffer, O. T. Didri, and E. Walterm. *Applied Interval Analysis*. Springer, London, 2001. DOI: 10.1007/978-1-4471-0249-6. 13

[3] T. J. Ross. *Fuzzy Logic with Engineering Applications*. John Wiley & Sons, New York, 2004. DOI: 10.1002/9781119994374. 13

[4] M. Hanss. *Applied Fuzzy Arithmetic: An Introduction with Engineering Applications*. Springer-Verlag, Berlin, 2005. DOI: 10.1007/b138914. 13

[5] R. E. Moore. *Interval Analysis*. Prentice Hall, Englewood Cliffs, NJ, 1966. 13

[6] S. Chakraverty. *Mathematics of Uncertainty Modeling in the Analysis of Engineering and Science Problems*. IGI Global Publication, 2014. DOI: 10.4018/978-1-4666-4991-0. 13

[7] S. Chakraverty, S. Tapaswini, and D. Behera. *Fuzzy Arbitrary Order System: Fuzzy Fractional Differential Equations and Applications*. John Wiley & Sons, 2016. DOI: 10.1002/9781119004233. 13

[8] S. Chakraverty, S. Tapaswini, and D. Behera. *Fuzzy Differential Equations and Applications for Engineers and Scientists*. Taylor & Francis Group, CRC Press, Boca Raton, FL, 2016. DOI: 10.1201/9781315372853.

[9] S. Chakraverty, D. M. Sahoo, and N. R. Mahato. *Concepts of Soft Computing: Fuzzy and ANN with Programming*. Springer, Singapore, 2019. DOI: 10.1007/978-981-13-7430-2. 13

[10] Y. Chalco-Cano, and H. Roman-Flores. On new solutions of fuzzy differential equations. *Chaos Solitons Fract*, 38:112–119, 2008. DOI: 10.1016/j.chaos.2006.10.043. 18, 19

[11] A. Khastan, J. J. Nieto, and R. Rodriguez-Lopez. Variation of constant formula for first order fuzzy differential equations. *Fuzzy Sets Systems*, 177:20–33, 2011. DOI: 10.1016/j.fss.2011.02.020. 18

[12] M. Mazandarania and A. V. Kamyad. Modified fractional Euler method for solving fuzzy fractional initial value problem, *Commun. Nonlinear Sci. Numer. Simulat.*, 18(1):12 21, 2013. DOI: 10.1016/j.cnsns.2012.06.008.

[13] B. Bede. Note on numerical solutions of fuzzy differential equations by predictor-corrector method. *Information Sciences*, 178:1917–1922, 2008. DOI: 10.1016/j.ins.2007.11.016. 21

[14] V. Lakshmikantham and R. N. Mohapatra. *Theory of Fuzzy Differential Equations and Applications*. Taylor & Francis, London, 2003. DOI: 10.1201/9780203011386.

CHAPTER 3

Fuzzy Fractional Differential Equations and Method of Solution

3.1 INTRODUCTION

In the previous chapter, we discussed preliminaries of fuzzy set theory and related definitions as well as notations which will be helpful for readers to understand the basics of fuzzy set theory. In this chapter, we discuss the theories regarding the fractional reduced differential transform method (FRDTM). Few numerical examples related to fractional differential equations are solved using the FRDTM. The same method has been extended to solve fuzzy fractional differential equations using the double parametric form.

Zhou, 1986 [1] first proposed differential transform method (DTM), and it was used to solve linear and nonlinear initial value problems arising in electric circuit analysis. Later, DTM was used to solve partial differential equations [18]. An analytical version of DTM [19] was then proposed which is named reduced differential transform method (RDTM). In this regard, RDTM is found to be effective and reliable for handling different linear and nonlinear partial differential equations and integral equations ([2]–[4]). In 2010, Keskin and Oturanc [5] applied this method to solve fractional differential equations with some other modifications. Later this method was used in fractional differential equations, which is known as FRDTM. This method has been successfully applied to solve various types of fractional partial differential equations ([6]–[8]), including higher-dimensional problems [9].

It is worth mentioning that FRDTM is a simple implementable semi-analytical technique and provides an approximate analytical solution for both linear, nonlinear, and fuzzy fractional differential equations. It does not use any discretization, transformation, perturbation, or any restrictive conditions. This method needs fewer computations in comparison with other perturbation methods.

3.2   DESCRIPTION OF FRACTIONAL REDUCED DIFFERENTIAL TRANSFORM METHOD

In order to understand the basic concept of FRDTM [6, 7, 10], first we recall and review the local fractional Taylor's theorems, and then, we extend it to FRDTM for fractional derivative.

Theorem 3.1   Local fractional Taylor's theorem.   *Suppose that $\zeta^{(k+1)\alpha} \in C_\alpha(a,b)$, for $k = 0, 1, \ldots, n$ and $0 < \alpha \le 1$, then we have*

$$\zeta(x) = \sum_{k=0}^{\infty} \zeta^{(k\alpha)}(0) \frac{(x - x_0)^{k\alpha}}{\Gamma(1 + k\alpha)}, \tag{3.1}$$

where $a < x_0 < x < b$, $\forall x \in (a,b)$ and $\zeta^{(k+1)\alpha}(x) = \underbrace{D_x^\alpha D_x^\alpha D_x^\alpha \ldots D_x^\alpha}_{k+1 \; times} \zeta(x).$

Definition 3.2   The fractional reduced differential transform $\zeta_k(x)$ of an analytic function $\zeta(x,t)$ is defined as

$$\zeta_k(x) = \frac{1}{\Gamma(1 + \alpha k)} \left(\frac{\partial^{k\alpha} \zeta(x,t)}{\partial t^{k\alpha}} \right)_{t=0} \qquad where \quad k = 0, 1, \ldots, n. \tag{3.2}$$

Definition 3.3   The fractional inverse differential transform of $\zeta_k(x)$ is defined as

$$\zeta(x,t) = \sum_{k=0}^{\infty} \zeta_k(x) t^{k\alpha}. \tag{3.3}$$

Using Eqs. (3.2) and (3.3), the following theorems of FRDTM are deduced.

Theorem 3.4   *Let $\psi(x,t), \xi(x,t)$ and $\zeta(x,t)$ are three analytical functions such that $\psi(x,t) = R_D^{-1}[\psi_k(x)]$, $\xi(x,t) = R_D^{-1}[\xi_k(x)]$ and $\zeta(x,t) = R_D^{-1}[\zeta_k(x)]$. Hence*

(i) *If $\psi(x,t) = c_1 \xi(x,t) \pm c_2 \zeta(x,t)$, then $\psi_k(x) = c_1 \xi_k(x) \pm c_2 \zeta_k(x)$, where c_1 and c_2 are constants.*

(ii) *If $\psi(x,t) = a\xi(x,t)$, then $\psi_k(x) = a\xi_k(x)$.*

(iii) *If $\psi(x,t) = \xi(x,t)\zeta(x,t)$, then $\psi_k(x) = \sum_{i=0}^{j} \xi_i(x)\zeta_{j-i}(x) = \sum_{i=0}^{j} \zeta_i(x)\xi_{j-i}(x).$*

(iv) *If $\psi(x,t) = \frac{\partial^m}{\partial x^m}\xi(x,t)$, then $\psi_k(x) = \frac{\partial^m}{\partial x^m}\xi_k(x).$*

(v) *If $\psi(x,t) = \frac{\partial^{n\alpha}}{\partial t^{n\alpha}}\xi(x,t)$, then $\psi_k(x) = \frac{\Gamma(1+(k+n)\alpha)}{(1+k\alpha)}\xi_{k+n}(x),$*

where R_D^{-1} denotes the inverse reduced differential transform operator.

Proof.   (i) From Eq. (3.2), we have

$$\psi_k(x) = \frac{1}{\Gamma(1+\alpha k)}\left(\frac{\partial^{k\alpha}\psi(x,t)}{\partial t^{k\alpha}}\right)_{t=0}$$

$$= \frac{1}{\Gamma(1+\alpha k)}\left(\frac{\partial^{k\alpha}}{\partial t^{k\alpha}}(c_1\xi(x,t)\pm c_2\zeta(x,t))\right)_{t=0}$$

$$= \frac{1}{\Gamma(1+\alpha k)}\left(c_1\frac{\partial^{k\alpha}\xi(x,t)}{\partial t^{k\alpha}}\pm c_2\frac{\partial^{k\alpha}\zeta(x,t)}{\partial t^{k\alpha}}\right)_{t=0}$$

$$= c_1\frac{1}{\Gamma(1+\alpha k)}\left(\frac{\partial^{k\alpha}\xi(x,t)}{\partial t^{k\alpha}}\right)_{t=0}\pm c_2\frac{1}{\Gamma(1+\alpha k)}\left(\frac{\partial^{k\alpha}\zeta(x,t)}{\partial t^{k\alpha}}\right)_{t=0}$$

$$\psi_k(x) = c_1\xi_k(x)\pm c_2\zeta_k(x). \tag{3.4}$$

(ii) Putting $\psi(x,t) = a\xi(x,t)$ in Eq. (3.2), we have

$$\psi_k(x) = \frac{1}{\Gamma(1+\alpha k)}\left(\frac{\partial^{k\alpha}\psi(x,t)}{\partial t^{k\alpha}}\right)_{t=0}$$

$$= \frac{1}{\Gamma(1+\alpha k)}\left(\frac{\partial^{k\alpha}}{\partial t^{k\alpha}}(a\xi(x,t))\right)_{t=0}$$

$$= a\frac{1}{\Gamma(1+\alpha k)}\left(\frac{\partial^{k\alpha}\xi(x,t)}{\partial t^{k\alpha}}\right)_{t=0} = a\xi_k(x). \tag{3.5}$$

(iii) From Eq. (3.3), we obtain

$$\psi(x,t) = \left(\sum_{k=0}^{\infty}\xi_k(x)t^{k\alpha}\right)\left(\sum_{k=0}^{\infty}\zeta_k(x)t^{k\alpha}\right)$$

$$= \left(\xi_0(x)+\xi_1(x)t^{\alpha}+\xi_2(x)t^{2\alpha}+\cdots\right)\left(\zeta_0(x)+\zeta_1(x)t^{\alpha}+\zeta_2(x)t^{2\alpha}+\cdots\right)$$

$$= \left(\begin{array}{l}\xi_0(x)\zeta_0(x)+(\xi_0(x)\zeta_1(x)+\xi_1(x)\zeta_0(x))t^{\alpha}+(\xi_0(x)\zeta_2(x)\\+\xi_1(x)\zeta_1(x)+\xi_2(x)\zeta_0(x))t^{2\alpha}+\cdots+(\xi_0(x)\zeta_k(x)+\xi_1(x)\zeta_{k-1}(x)\\+\cdots+\xi_{k-1}(x)\zeta_1(x)+\xi_k(x)\zeta_0(x))t^{2\alpha}\end{array}\right)$$

$$\psi_k(x) = \sum_{i=0}^{j}\xi_i(x)\zeta_{j-i}(x) = \sum_{i=0}^{j}\zeta_i(x)\xi_{j-i}(x). \tag{3.6}$$

(iv) From Eq. (3.2), we have

$$\psi_k(x) = \frac{1}{\Gamma(1+\alpha k)}\left(\frac{\partial^{k\alpha}\psi(x,t)}{\partial t^{k\alpha}}\right)_{t=0}.$$

Substituting $\psi(x,t) = \frac{\partial^m}{\partial x^m}\xi(x,t)$ in the above equation, we get

$$\psi_k(x) = \frac{1}{\Gamma(1+\alpha k)}\left(\frac{\partial^{k\alpha}}{\partial t^{k\alpha}}\left(\frac{\partial^m}{\partial x^m}\xi(x,t)\right)\right)_{t=0}$$

$$= \frac{1}{\Gamma(1+\alpha k)}\left(\frac{\partial^m}{\partial x^m}\left(\frac{\partial^{k\alpha}\xi(x,t)}{\partial t^{k\alpha}}\right)\right)_{t=t_0} = \frac{\partial^m \xi_k(x)}{\partial x^m}. \tag{3.7}$$

(v) We have

$$\psi_k(x) = \frac{1}{\Gamma(1+\alpha k)}\left(\frac{\partial^{k\alpha}\psi(x,t)}{\partial t^{k\alpha}}\right)_{t=0}$$

$$= \frac{1}{\Gamma(1+\alpha k)}\left(\frac{\partial^{k\alpha}}{\partial t^{k\alpha}}\left(\frac{\partial^{n\alpha}}{\partial t^{n\alpha}}\xi(x,t)\right)\right)_{t=0}$$

$$= \frac{1}{\Gamma(1+\alpha k)}\left(\frac{\partial^{(k+n)\alpha}}{\partial t^{(k+n)\alpha}}(\xi(x,t))\right)_{t=0}$$

$$\psi_k(x) = \frac{\Gamma(1+(k+n)\alpha)}{(1+k\alpha)}\xi_{k+n}(x). \tag{3.8}$$

To explain the implementation of FRDTM, the following equation in operator form is considered

$$D_t^{n\alpha}\psi(x,t) + R\psi(x,t) + N\psi(x,t) = f(x,t), n-1 < n\alpha \le n, \tag{3.9}$$

with initial conditions:

$$\psi^{(i)}(x,0) = g_i(x), \quad i = 0,1,\ldots,n-1, \tag{3.10}$$

where R and N are linear and nonlinear differential operators, respectively, and $f(x,t)$ is an inhomogeneous source term. Then, by using Theorem 3.4 (property (v)) and Eqs. (3.2), (3.9) reduces to

$$\frac{\Gamma(1+\alpha k + n\alpha)}{\Gamma(1+\alpha k)}\psi_{k+n}(x) = F_k(x) - R\psi_k(x) - N\psi_k(x), \quad \text{for} \quad k = 0,1,2,\ldots$$

$$\tag{3.11}$$

where $\psi_k(x)$ and $F_k(x)$ are the differential transformed form of $\psi(x,t)$ and $f(x,t)$, respectively.

Applying FRDTM on the initial conditions, we obtain

$$\psi_k(x) = g_k(x), \quad \text{for} \quad k = 0, 1, 2, \ldots, n-1. \tag{3.12}$$

Using Eqs. (3.11) and (3.12), $\psi_k(x)$ for $k = 0, 1, 2, 3, \ldots$ can be determined.

Applying the inverse transformation to $\{\psi_k(x)\}_{k=0}^{n}$, we get n-term approximate solution as

$$\psi_n(x,t) = \sum_{k=0}^{n} \psi_k(x)\, t^{\alpha k}. \tag{3.13}$$

So, the analytical result of Eq. (3.9) may be written as $\psi(x,t) = \lim_{n\to\infty} \psi_n(x,t)$.

$\square$

One may see the reference [14] for the convergence study of the present method. Interested authors may also see the references [6, 7, 10] for further details about the fractional reduced differential transform.

3.3   NUMERICAL EXAMPLES

Here, we apply the present method to solve one-dimensional heat-like fractional model in Example 3.5, nonlinear fractional advection equation in Example 3.6 and fuzzy fractional differential equation in Example 3.7.

3.3.1   TIME-FRACTIONAL LINEAR DIFFERENTIAL EQUATION

Example 3.5   Let us consider the one-dimensional heat-like model [11, 12] as

$$D_t^\alpha \psi(x,t) = \frac{1}{2}x^2 \psi_{xx}(x,t), \quad 0 < x < 1, \quad t > 0, \quad 0 < \alpha \le 1, \tag{3.14}$$

with the boundary conditions (BCs)

$$\psi(0,t) = 0, \quad \psi(1,t) = e^t \tag{3.15}$$

and initial condition (IC)

$$\psi(x,0) = x^2. \tag{3.16}$$

It may be noted that this is the fractional differential equation in the crisp environment.

Solution: By applying FRDTM to Eq. (3.14) and using Theorems 3.4 (iv) and (v), the following recurrence relation is obtained:

$$\frac{\Gamma\left(1+(k+1)\alpha\right)}{(1+k\alpha)}\psi_{k+1}(x) = \frac{1}{2}x^2\frac{\partial^2}{\partial x^2}\psi_k(x). \tag{3.17}$$

By using FRDTM to the initial condition Eq. (3.16), we get

$$\psi_0(x) = x^2. \tag{3.18}$$

Using Eq. (3.18) into Eq. (3.17), the following values of $\psi_k(x)$ for $k = 0, 1, 2, \ldots$ are obtained successively:

$$k = 0 : \psi_1(x) = \left(\frac{1}{2}x^2\frac{\partial^2}{\partial x^2}\psi_0(x)\right) = \frac{1}{\Gamma(1+\alpha)}\left(\frac{1}{2}2x^2\right) = \frac{x^2}{\Gamma(1+\alpha)}, \tag{3.19}$$

$$k = 1 : \psi_2(x) = \frac{\Gamma(1+\alpha)}{\Gamma(1+2\alpha)}\left(\frac{1}{2}x^2\frac{\partial^2}{\partial x^2}\psi_1(x)\right)$$

$$= \frac{\Gamma(1+\alpha)}{\Gamma(1+2\alpha)}\left(\frac{1}{2}x^2\frac{2}{\Gamma(1+\alpha)}\right) = \frac{x^2}{\Gamma(1+2\alpha)}, \tag{3.20}$$

$$k = 2 : \psi_3(x) = \frac{\Gamma(1+2\alpha)}{\Gamma(1+3\alpha)}\left(\frac{1}{2}x^2\frac{\partial^2}{\partial x^2}\psi_2(x)\right)$$

$$= \frac{\Gamma(1+2\alpha)}{\Gamma(1+3\alpha)}\left(\frac{1}{2}x^2\frac{2}{\Gamma(1+2\alpha)}\right) = \frac{x^2}{\Gamma(1+3\alpha)}, \tag{3.21}$$

and so on.

In the same manner, the rest of the components of the iteration formula Eq. (3.17) can be obtained. So, the solution of the Eq. (3.14) maybe written as

$$\psi(x,t) = \sum_{k=0}^{\infty}\psi_k(x)t^{\alpha k},$$

$$= \psi_0(x) + \psi_1(x)t^\alpha + \psi_2(x)t^{2\alpha} + \ldots,$$

$$= x^2 + \frac{x^2}{\Gamma(1+\alpha)} + \frac{x^2}{\Gamma(1+2\alpha)} + \frac{x^2}{\Gamma(1+3\alpha)} + \ldots = x^2E(t^\alpha), \tag{3.22}$$

which is same as the solution given in refs. [11, 12], where $E(t^\alpha)$ is called the MLF.

3.3.2   TIME-FRACTIONAL NONLINEAR DIFFERENTIAL EQUATION

Example 3.6   Consider the following nonlinear advection equation [13]

$$\frac{\partial^\alpha\psi}{\partial t^\alpha} + \psi\frac{\partial\psi}{\partial x} = 0, \quad 0 < \alpha \le 1 \tag{3.23}$$

with an IC:

$$\psi(x, 0) = -x. \tag{3.24}$$

Again, the above is a fractional differential equation in crisp environment.

Solution: Using FRDTM on the above two equations, we obtain the following recurrence relation:

$$\frac{\Gamma(1 + (k+1)\alpha)}{(1 + k\alpha)} \psi_{k+1}(x) = -\left(\sum_{i=0}^{k} \psi_i(x) \frac{\partial}{\partial x} \psi_{k-i}(x)\right), \\ \psi_0(x) = -x. \left.\begin{array}{c} \\ \\ \end{array}\right\} \tag{3.25}$$

On solving Eq. (3.25), we have

$$k = 0 : \psi_1(x) = \frac{\Gamma(1)}{\Gamma(1 + \alpha)}\left(-\psi_0(x)\frac{\partial}{\partial x}\psi_0(x)\right) = \frac{-1}{\Gamma(1 + \alpha)}(-x)(-1)$$

$$= \frac{-x}{\Gamma(1 + \alpha)}, \tag{3.26}$$

$$k = 1 : \psi_2(x) = \frac{-\Gamma(1 + \alpha)}{\Gamma(1 + 2\alpha)}\left(\psi_0(x)\frac{\partial}{\partial x}\psi_1(x) + \psi_1(x)\frac{\partial}{\partial x}\psi_0(x)\right)$$

$$= \frac{-\Gamma(1 + \alpha)}{\Gamma(1 + 2\alpha)}\left\{(-x)\left(\frac{-1}{\Gamma(1 + \alpha)}\right) + \left(\frac{-x}{\Gamma(1 + \alpha)}\right)(-1)\right\}$$

$$= \frac{-2x}{\Gamma(1 + 2\alpha)}, \tag{3.27}$$

$$k = 2 : \psi_3(x) = \frac{-\Gamma(1 + 2\alpha)}{\Gamma(1 + 3\alpha)}\left\{\psi_0\frac{\partial}{\partial x}\psi_2 + \psi_1\frac{\partial}{\partial x}\psi_1 + \psi_2\frac{\partial}{\partial x}\psi_0\right\}$$

$$= \frac{-\Gamma(1 + 2\alpha)}{\Gamma(1 + 3\alpha)}\left\{(-x)\left(\frac{-2}{\Gamma(1 + 2\alpha)}\right) + \left(\frac{-x}{\Gamma(1 + \alpha)}\right)\left(\frac{-1}{\Gamma(1 + \alpha)}\right)\right.$$

$$\left. + \left(\frac{-2x}{\Gamma(1 + 2\alpha)}\right)(-1)\right\}$$

$$= \frac{-\Gamma(1 + 2\alpha)}{\Gamma(1 + 3\alpha)}\left\{\frac{4x}{\Gamma(1 + 2\alpha)} + \frac{x}{(\Gamma(1 + \alpha))^2}\right\}$$

$$= \frac{-4x}{\Gamma(1 + 3\alpha)} + \frac{-\Gamma(1 + 2\alpha)}{\Gamma(1 + 3\alpha)}\frac{x}{(\Gamma(1 + \alpha))^2}, \tag{3.28}$$

and so on.

Similarly, the rest of the components $\psi_k(x)$ for $k = 3, 4, \ldots$ can be computed. So, the solution of the Eq. (3.18) maybe written as

$$
\begin{aligned}
\psi(x, t) &= \sum_{k=0}^{\infty} \psi_k(x) t^{\alpha k}, \\
&= \psi_0(x) + \psi_1(x) t^{\alpha} + \psi_2(x) t^{2\alpha} + \cdots, \\
&= -x + \frac{-x}{\Gamma(1 + \alpha)} t^{\alpha} + \frac{-2x}{\Gamma(1 + 2\alpha)} t^{2\alpha} + \frac{-4x}{\Gamma(1 + 3\alpha)} t^{3\alpha} \\
&\quad + \frac{-\Gamma(1 + 2\alpha)}{\Gamma(1 + 3\alpha)} \frac{x}{(\Gamma(1 + \alpha))^2} t^{3\alpha} + \cdots.
\end{aligned}
\tag{3.29}
$$

In particular at $\alpha = 1$, Eq. (3.29) reduces to

$$
\psi(x, t) = -x - xt - xt^2 - xt^3 - \ldots = -x\left(1 + t + t^2 + t^3 + \ldots\right) = \frac{-x}{1 - t}, \tag{3.30}
$$

which is same as the solution of ref. [13].

3.3.3　TIME-FRACTIONAL FUZZY DIFFERENTIAL EQUATION

Example 3.7　Let us now consider the following fuzzy fractional differential equation (which is our main aim)

$$
\frac{\partial^{\alpha} \tilde{\psi}(t; \gamma)}{\partial t^{\alpha}} = \tilde{\psi}(t; \gamma), \quad 0 < \alpha \le 1 \quad \text{and} \quad \gamma \in [0, 1], \tag{3.31}
$$

subject to a triangular fuzzy IC:

$$
\tilde{\psi}(0) = (0.8, \ 1, \ 1.2). \tag{3.32}
$$

It is worth mentioning here that the above IC is given in terms of a TFN. And so the fractional differential equation is now the fuzzy fractional differential equation.

Solution:　As per the single parametric form of the fuzzy number (as discussed in Chapter 2), Eq. (3.31) may now be written (in term of γ-cut) as

$$
\left[\frac{\partial^{\alpha} \underline{\psi}(t; \gamma)}{\partial t^{\alpha}}, \frac{\partial^{\alpha} \overline{\psi}(t; \gamma)}{\partial t^{\alpha}} \right] = \left[\underline{\psi}(t; \gamma), \overline{\psi}(t; \gamma) \right], \tag{3.33}
$$

subject to fuzzy initial condition:

$$
\left[\underline{\psi}(0; \gamma), \overline{\psi}(0; \gamma) \right] = [0.2\gamma + 0.8, \ 1.2 - 0.2\gamma]. \tag{3.34}
$$

Next, by using the DPF of the fuzzy number (as discussed in Chapter 2), Eqs. (3.33) and (3.34) may be expressed as

$$\beta \left(\frac{\partial^\alpha \overline{\psi}(t;\gamma)}{\partial t^\alpha} - \frac{\partial^\alpha \underline{\psi}(t;\gamma)}{\partial t^\alpha} \right) + \frac{\partial^\alpha \underline{\psi}(t;\gamma)}{\partial t^\alpha} = \beta \left(\overline{\psi}(t;\gamma) - \underline{\psi}(t;\gamma) \right) + \underline{\psi}(t;\gamma), \quad (3.35)$$

with fuzzy initial condition

$$\beta \left(\overline{\psi}(0;\gamma) - \underline{\psi}(0;\gamma) + \underline{\psi}(0;\gamma) \right) = \beta(0.4 - 0.4\gamma) + 1.2 - 0.2\gamma, \quad (3.36)$$

where $\gamma, \beta \in [0, 1]$.

It may again be noted that the two parameters γ and β control the fuzzy uncertainty of the problem.

Let us now denote

$$\beta \left(\frac{\partial^\alpha \overline{\psi}(t;\gamma)}{\partial t^\alpha} - \frac{\partial^\alpha \underline{\psi}(t;\gamma)}{\partial t^\alpha} \right) + \frac{\partial^\alpha \underline{\psi}(t;\gamma)}{\partial t^\alpha} = \frac{\partial^\alpha \tilde{\psi}(t;\gamma,\beta)}{\partial t^\alpha},$$

$$\beta \left(\overline{\psi}(t;\gamma) - \underline{\psi}(t;\gamma) \right) + \underline{\psi}(t;\gamma) = \tilde{\psi}(t;\gamma,\beta),$$

$$\beta \left(\overline{\psi}(0;\gamma) - \underline{\psi}(0;\gamma) + \underline{\psi}(0;\gamma) \right) = \tilde{\psi}(0;\gamma,\beta).$$

Substituting these values in Eqs. (3.35) and (3.36), we get

$$\frac{\partial^\alpha \tilde{\psi}(t;\gamma,\beta)}{\partial t^\alpha} = \tilde{\psi}(t;\gamma,\beta), \quad (3.37)$$

$$\tilde{\psi}(0;\gamma,\beta) = \beta(0.4 - 0.4\gamma) + 1.2 - 0.2\gamma = \eta \quad (\text{say}). \quad (3.38)$$

On solving the above fractional parametrized differential equation (Eqs. (3.37) and (3.38)), one may obtain the solution in terms of $\tilde{\psi}(t;\gamma,\beta)$. In order to obtain the lower- and upper-bound solutions of Eqs. (3.37) and (3.38), we need to put $\beta = 0$ and $\beta = 1$, respectively. Mathematically, this may be written as

$$\tilde{\psi}(t;\gamma,0) = \underline{\psi}(t,\gamma) \quad \text{and} \quad \tilde{\psi}(t;\gamma,1) = \overline{\psi}(t,\gamma).$$

Applying FRDTM to Eqs. (3.37)–(3.38) and using Theorem 3.4 (v), the following recurrence relation is obtained:

$$\left. \begin{array}{c} \dfrac{\Gamma(1 + (k+1)\alpha)}{(1 + k\alpha)} \tilde{\psi}_{k+1}(\gamma,\beta) = \tilde{\psi}_k(\gamma,\beta), \\[2mm] \tilde{\psi}_0(\gamma,\beta) = \eta. \end{array} \right\} \quad (3.39)$$

On solving Eq. (3.39), the following values of $\psi_k(x)$ for $k = 1, 2, \ldots$ are obtained successively:

$$k = 0 : \tilde{\psi}_1(\gamma, \beta) = \frac{\Gamma(1)}{\Gamma(1+\alpha)}(\tilde{\psi}_0(\gamma, \beta)) = \frac{\eta}{\Gamma(1+\alpha)}, \tag{3.40}$$

$$k = 1 : \tilde{\psi}_2(\gamma, \beta) = \frac{\Gamma(1+\alpha)}{\Gamma(1+2\alpha)}(\tilde{\psi}_1(\gamma, \beta))$$

$$= \frac{\Gamma(1+\alpha)}{\Gamma(1+2\alpha)}\left(\frac{\eta}{\Gamma(1+\alpha)}\right) = \frac{\eta}{\Gamma(1+2\alpha)}, \tag{3.41}$$

$$k = 2 : \tilde{\psi}_3(\gamma, \beta) = \frac{\Gamma(1+2\alpha)}{\Gamma(1+3\alpha)}(\tilde{\psi}_2(\gamma, \beta))$$

$$= \frac{\Gamma(1+2\alpha)}{\Gamma(1+3\alpha)}\left(\frac{\eta}{\Gamma(1+2\alpha)}\right) = \frac{\eta}{\Gamma(1+3\alpha)}, \tag{3.42}$$

and so on.

Likewise, the rest of the components of the iteration formula Eq. (3.39) can be obtained. So, the solution of the Eq. (3.39) maybe written as

$$\tilde{\psi}(t; \gamma, \beta) = \sum_{k=0}^{\infty} \tilde{\psi}_k(\gamma, \beta) t^{\alpha k},$$

$$= \tilde{\psi}_0(\gamma, \beta) + \tilde{\psi}_1(\gamma, \beta) t^{\alpha} + \tilde{\psi}_2(\gamma, \beta) t^{2\alpha} + \ldots,$$

$$= \eta + \frac{\eta}{\Gamma(1+\alpha)} + \frac{\eta}{\Gamma(1+2\alpha)} + \frac{\eta}{\Gamma(1+3\alpha)} + \ldots = \eta E(t^{\alpha}). \tag{3.43}$$

So, the solution of Eqs. (3.31) and (3.32) may be expressed as

$$\tilde{\psi}(t; \gamma, \beta) = [\beta(0.4 - 0.4\gamma) + 1.2 - 0.2\gamma] E(t^{\alpha}), \tag{3.44}$$

where $E(t^{\alpha})$ is called the Mittag–Leffler function.

The lower and upper bounds of the solution of Eqs. (3.31)–(3.32) are $\tilde{\psi}(t; \gamma, 0) = (1.2 - 0.2\gamma) E(t^{\alpha})$ and $\tilde{\psi}(t; \gamma, 1) = (1.6 - 0.6\gamma) E(t^{\alpha})$, respectively.

It is interesting to note that if we substitute $\gamma = 1$ in Eq. (3.44), then we obtain $\tilde{\psi}(t; 1, \beta) = E(t^{\alpha})$, which is the solution of Eqs. (3.31)–(3.32) in crisp form.

In Example 3.7, we have used triangular fuzzy initial condition, but one may use initial condition as a different fuzzy number like TrFN, GFN, and so on as per the need or application. The test problems confirm that the FRDTM is an efficient method for solving linear/nonlinear/fuzzy fractional differential equations. The series usually converges with an increase in the number of terms, but one may not always expect the compact form of the solution.

3.4 REFERENCES

[1] J. Zhou. *Differential Transformation and its Application for Electrical Circuits*. Huazhong University Press, Wuhan, China, 1986. 25

[2] R. Abazari and A. Kilicman. Numerical study of two-dimensional Volterra integral equations by RDTM and comparison with DTM. *Abstract and Applied Analysis*, 10:929478, 2013. DOI: 10.1155/2013/929478. 25

[3] R. Abazari and B. Soltanalizadeh. Reduced differential transform method and its application on Kawahara equations. *Thai Journal of Mathematics*, 11:199–216, 2013.

[4] A. Saravanan and N. Magesh. A comparison between the reduced differential transform method and the Adomian decomposition method for the Newell–Whitehead–Segel equation. *Journal of the Egyptian Mathematical Society*, 21:259–265, 2013. DOI: 10.1016/j.joems.2013.03.004. 25

[5] Y. Keskin and G. Oturanc. Reduced differential transform method: A new approach to fractional partial differential equations. *Nonlinear Science Letters A*, 1:207–218, 2010. 25

[6] R. M. Jena, S. Chakraverty, and D. Baleanu. On the solution of imprecisely defined nonlinear time-fractional dynamical model of marriage. *Mathematics*, 7:689, 2019. DOI: 10.3390/math7080689. 25, 26, 29

[7] R. M. Jena, S. Chakraverty, and D. Baleanu. On new solutions of time-fractional wave equations arising in Shallow water wave propagation. *Mathemathics*, 7:722, 2019. DOI: 10.3390/math7080722. 26, 29

[8] A. Saravanan and N. Magesh. An efficient computational technique for solving the Fokker–Planck equation with space and time fractional derivatives. *Journal of King Saud University-Science*, 28:160–166, 2016. DOI: 10.1016/j.jksus.2015.01.003. 25

[9] M. S. Rawashdeh. An efficient approach for time–fractional damped burger and time–sharma–tasso–olver equations using the FRDTM. *Applied Mathematics and Information Sciences*, 9(3):1239–1246, 2015. 25

[10] H. Jafari, H. K. Jassim, S. P. Moshokoa, V. M. Ariyan, and F. Tchier. Reduced differential transform method for partial differential equations within local fractional derivative operators. *Advances in Mechanical Engineering*, 8(4):1–6, 2016. DOI: 10.1177/1687814016633013. 26, 29

[11] T. Özis and D. Agirseven. He's homotopy perturbation method for solving heat-like and wave-like equations, with variable coefficients. *Physics Letters A*, 372:5944–50, 2008. DOI: 10.1016/j.physleta.2008.07.060. 29, 30

[12] A. Sadighi, D. D. Ganji, M. Gorji, and N. Tolou. Numerical simulation of heat-like models with variable coefficients by the variational iteration method. *Journal of Physics Conference Series*, 96:012083, 2008. DOI: 10.1088/1742-6596/96/1/012083. 29, 30

[13] A. M. Wazwaz. A comparison between the variational iteration method and Adomian decomposition method. *Journal of Computational and Applied Mathematics*, 207:129–136, 2007. DOI: 10.1016/j.cam.2006.07.018. 30, 32

[14] A. Al-Saif and A. Harfash. A comparison between the reduced differential transform method and perturbation-iteration algorithm for solving two-dimensional unsteady incompressible Navier–Stokes equations. *Journal of Applied Mathematics and Physics*, 6:2518–2543, 2018. DOI: 10.4236/jamp.2018.612211. 29

[15] S. Chakraverty, S. Tapaswini, and D. Behera. *Fuzzy Arbitrary Order System: Fuzzy Fractional Differential Equations and Applications*. John Wiley & Sons, 2016. DOI: 10.1002/9781119004233.

[16] S. Chakraverty, S. Tapaswini, and D. Behera. *Fuzzy Differential Equations and Applications for Engineers and Scientists*. Taylor & Francis Group, CRC Press, Boca Raton, FL, 2016. DOI: 10.1201/9781315372853.

[17] S. Chakraverty, D. M. Sahoo, and N. R. Mahato. *Concepts of Soft Computing: Fuzzy and ANN with Programming*. Springer, Singapore, 2019. DOI: 10.1007/978-981-13-7430-2.

[18] Z. Odibat and S. Momani. A generalized differential transform method for linear partial differential equations of fractional order. *Applied Mathematics Letters*, 21:194–199, 2008. DOI: 10.1016/j.aml.2007.02.022. 25

[19] Y. Keskinand and G. Oturanç. Reduced differential transform method for partial differential equations. *International Journal of Nonlinear Sciences and Numerical Simulation*, 10(6):741–750, 2009. 25

CHAPTER 4

Imprecisely Defined Time-Fractional Model of Cancer Chemotherapy Effect

4.1 INTRODUCTION

In recent years, tumor growth and its treatment have been studied by different researchers using a variety of mathematical models. These mathematical models may provide us an analytical framework about the stages of tumor growth. If a person neglects the treatment of the tumor, then this tumor may be converted into a cancer cell. Cancer is a neoplastic disease, where abnormal growth of the cells is reproduced every day. One in eight human deaths is caused by cancer. The cancer treatment totally depends on the type and stages of cancer, and the treatment involves one or more of the following components: surgery, chemotherapy, radiation therapy, and so on. Surgery and radiation therapy may damage or kill the cancer cells in some infected areas, but chemotherapy treatment works throughout the whole body. In chemotherapy treatment, the doctors directly inject to cancer cells or through the bloodstream to make the cancer cells, frail and kill them. Over the past few decades, researchers have developed different mathematical modeling of the tumor-immune interaction [1–4]. Louzoun et al. [7] developed a mathematical model of pancreatic cancer which showed that drugs can suppress the cancer growth effectively only if the immune-induced cancer-cell death lies within a particular range. Weekes et al. [8] established a mathematical model showing that the exponential growth rate is identical to the growth rate of the cancer stem-cell compartment. Usman et al. [9] introduced a model of three reaction-diffusion equations that represent in vitro inhibition of cancer cell mutation. A system of five ordinary differential equations that consider population dynamics among cancer stem cells, tumor cells, and healthy cells was proposed by Abernathy et al. [10].

In this regard, fractional models are a recent development, and the biological phenomena modeled through derivatives of fractional order provide the information not only for the present state but also for past states [11]. Fractional derivative considers the system with memory, hereditary properties, and non-local behaviors. These properties are important for portraying the problems that arise in various types of science and engineering [5–7, 12–16]. As regards these problems, mathematical models involving fractional differential equations have been proven to be precious in understanding the nature of the tumor-immune system and how cancerous cells and host immune cells interact and evolve. Researchers also developed fractional-order model to study the diffusion equation to predict the effect of chemotherapy on cancer cells, the rate of change of cancerous blood cells in chronic myeloid leukemia, and brain tumor growth [17–19].

The primary aim of this chapter is to obtain the solution of an imprecisely defined time-fractional model of cancer chemotherapy effect using FRDTM. Here, the parameters involved in the titled problem are taken as fuzzy triangular numbers.

4.2   MATHEMATICAL MODEL

It is essential to understand biological problems through establishing mathematical models and analyzing their dynamical behaviors. In the current framework, we have considered the system of time-fractional differential equations describing the cancer chemotherapy effects in the Caputo sense, which consist of reaction and diffusion moment of the normal cells (N), tumor cells (T), immune cells (I), and chemotherapeutic drug (U) [5, 6, 20]

$$
\begin{cases}
\dfrac{\partial^\alpha N}{\partial t^\alpha} = \lambda_1 N - \omega_1 N^2 - c_4 TN + a_3 e^{-U} N + D_N \dfrac{\partial^2 N}{\partial x^2}, \\[2mm]
\dfrac{\partial^\alpha T}{\partial t^\alpha} = \lambda_2 T - \omega_2 T^2 - c_2 IT - c_3 TN + a_2 e^{-U} T + D_T \dfrac{\partial^2 T}{\partial x^2}, & 0 < \alpha \le 1, \\[2mm]
\dfrac{\partial^\alpha I}{\partial t^\alpha} = \varepsilon + \dfrac{\rho IT}{\mu + T} - c_1 IT - \lambda_3 I + a_1 e^{-U} I + D_I \dfrac{\partial^2 I}{\partial x^2}, & -2 \le x \le 2, \\[2mm]
\dfrac{\partial^\alpha U}{\partial t^\alpha} = v(t) - d_2 U + D_U \dfrac{\partial^2 U}{\partial x^2},
\end{cases}
\tag{4.1}
$$

subject to initial conditions:

$$
N(x,0) = 0.2 e^{-2x^2}, \quad T(x,0) = 1 - 0.75 \sec h(x),
$$
$$
I(x,0) = 0.375 - 0.235 \sec h^2(x), \quad U(x,0) = \sec h(x),
\tag{4.2}
$$

where

$$
\lambda_1 = r_2 - a_3, \quad \lambda_2 = r_1 - a_2, \quad \lambda_3 = d_1 - a_1, \quad \omega_1 = r_2 b_2, \quad \omega_2 = r_1 b_1. \tag{4.3}
$$

Table 4.1 shows the detailed description and values of the parameters in the governing Eq. (4.1) given in various studies [5, 6].

In Eq. (4.1), N, T, I, and U describe the host cells, which are tumor cells in a tumor; the immune cells; and the chemotherapeutic drug, respectively. The external rate of source for immune cells is represented by the parameter ε. The defense mechanism of the body in response to the presence of tumor cells is given by [5, 6] $\frac{\rho IT}{\mu + T}$. In immune cells, d_1 and d_2 indicate the per capita death and diminishing rate of the drug. Moreover, the range of some constraints are given as [21] $0 \le a_i \le 0.5, a_3 \le a_1 \le a_2$, and $b_2 \le b_1$ and c_1, c_2, c_3, and c_4 are the positive constants. Similarly, the bound for the rate of the immune source is $0 \le \varepsilon \le 0.5$. Further, the parameter

Table 4.1: Explanation of the constraints involved in Eq. (4.1)

Parameters	Descriptions	Values [5,6]
a_1, a_2, a_3	Fractional cell kill	0.2, 0.3, 0.1
b_1, b_2	Carrying capacity	1, 0.81
c_1, c_2, c_3, c_4	Competition term	1, 0.55, 0.9, 1
d_1, d_2	Death rate	0.2, 1
r_1, r_2	Per capita growth rate	1.1, 1
ε	Immune source rate	0.33
μ	Immune threshold rate	0.3
ρ	Immune response rate	0.2
D_N, D_T, D_I, D_U	Diffusion coefficient for normal, tumor, immune system cells, and the chemotherapeutic drug	0.001, 0.001, 0.001, 0.001

$v(t)$ represents the dosage of the drugs in each section of the treatment which is defined as follows:

$$v(t) = \begin{cases} 1 & \text{for} \quad (i-1)\xi < t < (i-1)\xi + \tau, \\ 0 & \text{for other,} \end{cases} \tag{4.4}$$

where ξ is the interval, τ is the duration, and $i = 1, 2, 3$. In the present model, initial conditions and tumor cells are considered under the assumption that the tumor cells are huge in size and the number of cells is 10^{11}.

4.3   MATHEMATICAL MODEL WITH FUZZY PARAMETERS

First, the time-fractional model of cancer chemotherapy effect is transformed into interval-based fuzzy fractional differential equations using gamma cut. Then, using the double parametric form, the interval-based fuzzy fractional differential equations are converted to a parametric form of fractional differential equations. Next, by applying FRDTM to the parametric form of differential equation, we obtain the fuzzy solution of the model. In this model, we have taken the parameter b_1, b_2, d_1, and d_2 as uncertain viz. in term of TFN, which are considered as follows:

$$\tilde{b}_1 = (0.9, 1, 1.1), \quad \tilde{b}_2 = (0.71, 0.81, 0.91),$$
$$\tilde{d}_1 = (0.1, 0.2, 0.3), \quad \text{and} \quad \tilde{d}_2 = (0.9, 1, 1.1). \tag{4.5}$$

Now, by substituting Eqs. (4.3) and (4.5) into Eq. (4.1), the following imprecisely defined time-fractional model of cancer chemotherapy effect is obtained:

$$
\begin{cases}
\dfrac{\partial^\alpha \tilde{N}}{\partial t^\alpha} = \lambda_1 \tilde{N} - r_2\,(0.71,\,0.81,\,0.91)\,\tilde{N}^2 - c_4 \tilde{T}\tilde{N} + a_3 e^{-\tilde{U}}\tilde{N} + D_N \dfrac{\partial^2 \tilde{N}}{\partial x^2}, \\[2mm]
\dfrac{\partial^\alpha \tilde{T}}{\partial t^\alpha} = \lambda_2 \tilde{T} - r_1\,(0.9,\,1,\,1.1)\,\tilde{T}^2 - c_2 \tilde{I}\tilde{T} - c_3 \tilde{T}\tilde{N} + a_2 e^{-\tilde{U}}\tilde{T} + D_T \dfrac{\partial^2 \tilde{T}}{\partial x^2}, \quad 0 < \alpha \le 1 \\[2mm]
\dfrac{\partial^\alpha \tilde{I}}{\partial t^\alpha} = \varepsilon + \dfrac{\rho \tilde{I}\tilde{T}}{\mu + \tilde{T}} - c_1 \tilde{I}\tilde{T} - [(0.1,\,0.2,\,0.3) - a_1]\,\tilde{I} + a_1 e^{-\tilde{U}}\tilde{I} + D_I \dfrac{\partial^2 \tilde{I}}{\partial x^2}, \quad -2 \le x \le 2 \\[2mm]
\dfrac{\partial^\alpha \tilde{U}}{\partial t^\alpha} = v\,(t) - (0.9,\,1,\,1.1)\,\tilde{U} + D_U \dfrac{\partial^2 \tilde{U}}{\partial x^2},
\end{cases}
$$

$$(4.6)$$

subject to initial condition Eq. (4.2). Here we have taken $v\,(t) = 1$.

Using γ-cut, Eq. (4.6) may be written as

$$
\begin{cases}
\left[\dfrac{\partial^\alpha \underline{N}(x,t\,;\gamma)}{\partial t^\alpha},\,\dfrac{\partial^\alpha \overline{N}(x,t\,;\gamma)}{\partial t^\alpha}\right] = \lambda_1 \left[\underline{N}\,(x,t\,;\gamma),\,\overline{N}\,(x,t\,;\gamma)\right] - r_2[0.1\gamma + 0.71,\,-0.1\gamma + 0.91] \\[2mm]
\left[\underline{N}\,(x,t\,;\gamma),\,\overline{N}\,(x,t\,;\gamma)\right]^2 - c_4 \left[\underline{T}\,(x,t\,;\gamma),\,\overline{T}\,(x,t\,;\gamma)\right]\left[\underline{N}\,(x,t\,;\gamma),\,\overline{N}\,(x,t\,;\gamma)\right] \\[2mm]
\quad + a_3 e^{-[\underline{u}(x,t\,;\gamma),\overline{u}(x,t\,;\gamma)]} \\[2mm]
\left[\underline{N}\,(x,t\,;\gamma),\,\overline{N}\,(x,t\,;\gamma)\right] + D_N \left[\dfrac{\partial^2 \underline{N}(x,t\,;\gamma)}{\partial x^2},\,\dfrac{\partial^2 \overline{N}(x,t\,;\gamma)}{\partial x^2}\right], \quad 0 < \alpha \le 1, \\[2mm]
\left[\dfrac{\partial^\alpha \underline{T}(x,t\,;\gamma)}{\partial t^\alpha},\,\dfrac{\partial^\alpha \overline{T}(x,t\,;\gamma)}{\partial t^\alpha}\right] = \lambda_2 \left[\underline{T}\,(x,t\,;\gamma),\,\overline{T}\,(x,t\,;\gamma)\right] - r_1\,[0.1\gamma + 0.9,\,-0.1\gamma + 1.1] \\[2mm]
\left[\underline{T}\,(x,t\,;\gamma),\,\overline{T}\,(x,t\,;\gamma)\right]^2 - c_2 \left[\underline{I}\,(x,t\,;\gamma),\,\overline{I}\,(x,t\,;\gamma)\right]\left[\underline{T}\,(x,t\,;\gamma),\,\overline{T}\,(x,t\,;\gamma)\right] \\[2mm]
\quad - c_3 \left[\underline{T}\,(x,t\,;\gamma),\,\overline{T}\,(x,t\,;\gamma)\right] \\[2mm]
\left[\underline{N}\,(x,t\,;\gamma),\,\overline{N}\,(x,t\,;\gamma)\right] + a_2 e^{-[\underline{u}(x,t\,;\gamma),\overline{u}(x,t\,;\gamma)]}\left[\underline{T}\,(x,t\,;\gamma),\,\overline{T}\,(x,t\,;\gamma)\right] \\[2mm]
\quad + D_T \left[\dfrac{\partial^2 \underline{T}(x,t\,;\gamma)}{\partial x^2},\,\dfrac{\partial^2 \overline{T}(x,t\,;\gamma)}{\partial x^2}\right], \\[2mm]
\left[\dfrac{\partial^\alpha \underline{I}(x,t\,;\gamma)}{\partial t^\alpha},\,\dfrac{\partial^\alpha \overline{I}(x,t\,;\gamma)}{\partial t^\alpha}\right] = \varepsilon + \dfrac{\rho\left[\underline{I}(x,t\,;\gamma),\overline{I}(x,t\,;\gamma)\right]\left[\underline{T}(x,t\,;\gamma),\overline{T}(x,t\,;\gamma)\right]}{\mu + \left[\underline{T}(x,t\,;\gamma),\overline{T}(x,t\,;\gamma)\right]} \\[2mm]
\quad - c_1 \left[\underline{I}\,(x,t\,;\gamma),\,\overline{I}\,(x,t\,;\gamma)\right] \\[2mm]
\left[\underline{T}\,(x,t\,;\gamma),\,\overline{T}\,(x,t\,;\gamma)\right] - \{[0.1\gamma + 0.1,\,-0.1\gamma + 0.3] - a_1\}\left[\underline{I}\,(x,t\,;\gamma),\,\overline{I}\,(x,t\,;\gamma)\right] \\[2mm]
\quad + a_1 e^{-[\underline{u}(x,t\,;\gamma),\overline{u}(x,t\,;\gamma)]} \\[2mm]
\left[\underline{I}\,(x,t\,;\gamma),\,\overline{I}\,(x,t\,;\gamma)\right] + D_I \left[\dfrac{\partial^2 \underline{I}(x,t\,;\gamma)}{\partial x^2},\,\dfrac{\partial^2 \overline{I}(x,t\,;\gamma)}{\partial x^2}\right], \\[2mm]
\left[\dfrac{\partial^\alpha \underline{U}(x,t\,;\gamma)}{\partial t^\alpha},\,\dfrac{\partial^\alpha \overline{U}(x,t\,;\gamma)}{\partial t^\alpha}\right] = 1 - [0.1\gamma + 0.9,\,-0.1\gamma + 1.1]\left[\underline{U}\,(x,t\,;\gamma),\,\overline{U}\,(x,t\,;\gamma)\right] \\[2mm]
\quad + D_U \left[\dfrac{\partial^2 \underline{U}(x,t\,;\gamma)}{\partial x^2},\,\dfrac{\partial^2 \overline{U}(x,t\,;\gamma)}{\partial x^2}\right], \quad -2 \le x \le 2.
\end{cases}
$$

$$(4.7)$$

By applying the double parametric form to Eq. (4.7), the following expressions in the parametric form are obtained:

$$
\begin{cases}
\left\{\beta\left(\frac{\partial^\alpha \overline{N}}{\partial t^\alpha} - \frac{\partial^\alpha \underline{N}}{\partial t^\alpha}\right) + \frac{\partial^\alpha \underline{N}}{\partial t^\alpha}\right\} = \lambda_1 \left\{\beta\left(\overline{N} - \underline{N}\right) + \underline{N}\right\} - r_2 \left\{\beta\left(0.2 - 0.2\gamma\right) + 0.1\gamma + 0.71\right\} \\[2mm]
\left\{\beta\left(\overline{N} - \underline{N}\right) + \underline{N}\right\}^2 - c_4 \left\{\beta\left(\overline{T} - \underline{T}\right) + \underline{T}\right\}\left\{\beta\left(\overline{N} - \underline{N}\right) + \underline{N}\right\} + a_3 e^{-\left\{\beta\left(\overline{U} - \underline{U}\right) + \underline{U}\right\}} \\[2mm]
\left\{\beta\left(\overline{N} - \underline{N}\right) + \underline{N}\right\} + D_N \left\{\beta\left(\frac{\partial^2 \overline{N}}{\partial x^2} - \frac{\partial^2 \underline{N}}{\partial x^2}\right) + \frac{\partial^2 \underline{N}}{\partial x^2}\right\}, \quad 0 < \alpha \le 1, \\[2mm]
\left\{\beta\left(\frac{\partial^\alpha \overline{T}}{\partial t^\alpha} - \frac{\partial^\alpha \underline{T}}{\partial t^\alpha}\right) + \frac{\partial^\alpha \underline{T}}{\partial t^\alpha}\right\} = \lambda_2 \left\{\beta\left(\overline{T} - \underline{T}\right) + \underline{T}\right\} - r_1 \left\{\beta\left(0.2 - 0.2\gamma\right) + 0.1\gamma + 0.9\right\} \\[2mm]
\left\{\beta\left(\overline{T} - \underline{T}\right) + \underline{T}\right\}^2 - c_2 \left\{\beta\left(\overline{I} - \underline{I}\right) + \underline{I}\right\}\left\{\beta\left(\overline{T} - \underline{T}\right) + \underline{T}\right\} - c_3 \left\{\beta\left(\overline{T} - \underline{T}\right) + \underline{T}\right\} \\[2mm]
\left\{\beta\left(\overline{N} - \underline{N}\right) + \underline{N}\right\} + a_2 e^{-\left\{\beta\left(\overline{U} - \underline{U}\right) + \underline{U}\right\}} \left\{\beta\left(\overline{T} - \underline{T}\right) + \underline{T}\right\} \\[2mm]
\qquad + D_T \left\{\beta\left(\frac{\partial^2 \overline{T}}{\partial x^2} - \frac{\partial^2 \underline{T}}{\partial x^2}\right) + \frac{\partial^2 \underline{T}}{\partial x^2}\right\}, \\[2mm]
\left\{\beta\left(\frac{\partial^\alpha \overline{I}}{\partial t^\alpha} - \frac{\partial^\alpha \underline{I}}{\partial t^\alpha}\right) + \frac{\partial^\alpha \underline{I}}{\partial t^\alpha}\right\} = \varepsilon + \frac{\rho\left\{\beta\left(\overline{I} - \underline{I}\right) + \underline{I}\right\}\left\{\beta\left(\overline{T} - \underline{T}\right) + \underline{T}\right\}}{\mu + \left\{\beta\left(\overline{T} - \underline{T}\right) + \underline{T}\right\}} - c_1 \left\{\beta\left(\overline{I} - \underline{I}\right) + \underline{I}\right\} \\[2mm]
\left\{\beta\left(\overline{T} - \underline{T}\right) + \underline{T}\right\} - \left\{\left(\beta\left(0.2 - 0.2\gamma\right) + 0.1\gamma + 0.1\right) - a_1\right\}\left\{\beta\left(\overline{I} - \underline{I}\right) + \underline{I}\right\} \\[2mm]
\qquad + a_1 e^{-\left\{\beta\left(\overline{U} - \underline{U}\right) + \underline{U}\right\}} \\[2mm]
\left\{\beta\left(\overline{I} - \underline{I}\right) + \underline{I}\right\} + D_I \left\{\beta\left(\frac{\partial^2 \overline{I}}{\partial x^2} - \frac{\partial^2 \underline{I}}{\partial x^2}\right) + \frac{\partial^2 \underline{I}}{\partial x^2}\right\}, \\[2mm]
\left\{\beta\left(\frac{\partial^\alpha \overline{U}}{\partial t^\alpha} - \frac{\partial^\alpha \underline{U}}{\partial t^\alpha}\right) + \frac{\partial^\alpha \underline{U}}{\partial t^\alpha}\right\} = 1 - \left\{\beta\left(0.2 - 0.2\gamma\right) + 0.1\gamma + 0.9\right\}\left\{\beta\left(\overline{U} - \underline{U}\right) + \underline{U}\right\} \\[2mm]
\qquad + D_U \left\{\beta\left(\frac{\partial^2 \overline{U}}{\partial x^2} - \frac{\partial^2 \underline{U}}{\partial x^2}\right) + \frac{\partial^2 \underline{U}}{\partial x^2}\right\}, \quad -2 \le x \le 2.
\end{cases}
$$

$$(4.8)$$

Let us denote

$$
\left\{\beta\left(\frac{\partial^\alpha \overline{N}}{\partial t^\alpha} - \frac{\partial^\alpha \underline{N}}{\partial t^\alpha}\right) + \frac{\partial^\alpha \underline{N}}{\partial t^\alpha}\right\} = \frac{\partial^\alpha \tilde{N}\left(x, t; \gamma\right)}{\partial t^\alpha},
$$

$$
\left\{\beta\left(\frac{\partial^\alpha \overline{T}}{\partial t^\alpha} - \frac{\partial^\alpha \underline{T}}{\partial t^\alpha}\right) + \frac{\partial^\alpha \underline{T}}{\partial t^\alpha}\right\} = \frac{\partial^\alpha \tilde{T}\left(x, t; \gamma\right)}{\partial t^\alpha},
$$

$$(4.9)$$

$$
\left\{\beta\left(\frac{\partial^\alpha \overline{I}}{\partial t^\alpha} - \frac{\partial^\alpha \underline{I}}{\partial t^\alpha}\right) + \frac{\partial^\alpha \underline{I}}{\partial t^\alpha}\right\} = \frac{\partial^\alpha \tilde{I}\left(x, t; \gamma\right)}{\partial t^\alpha},
$$

$$
\left\{\beta\left(\frac{\partial^\alpha \overline{U}}{\partial t^\alpha} - \frac{\partial^\alpha \underline{U}}{\partial t^\alpha}\right) + \frac{\partial^\alpha \underline{U}}{\partial t^\alpha}\right\} = \frac{\partial^\alpha \tilde{U}\left(x, t; \gamma\right)}{\partial t^\alpha},
$$

$$(4.10)$$

$$\{\beta\left(\overline{N}-\underline{N}\right)+\underline{N}\}=\tilde{N}\left(x,t\,;\gamma\right),$$
$$\{\beta\left(\overline{T}-\underline{T}\right)+\underline{T}\}=\tilde{T}\left(x,t\,;\gamma\right), \qquad (4.11)$$
$$\{\beta\left(\overline{I}-\underline{I}\right)+\underline{I}\}=\tilde{I}\left(x,t\,;\gamma\right),$$

$$\{\beta\left(\overline{U}-\underline{U}\right)+\underline{U}\},\quad \tilde{U}\left(x,t\,;\gamma\right),\quad \left\{\beta\left(\frac{\partial^2 \overline{N}}{\partial x^2}-\frac{\partial^2 \underline{N}}{\partial x^2}\right)+\frac{\partial^2 \underline{N}}{\partial x^2}\right\}=\frac{\partial^2 \tilde{N}\left(x,t\,;\gamma\right)}{\partial x^2}, \qquad (4.12)$$

$$\left\{\beta\left(\frac{\partial^2 \overline{T}}{\partial x^2}-\frac{\partial^2 \underline{T}}{\partial x^2}\right)+\frac{\partial^2 \underline{T}}{\partial x^2}\right\}=\frac{\partial^2 \tilde{T}\left(x,t\,;\gamma\right)}{\partial x^2},$$

$$\left\{\beta\left(\frac{\partial^2 \overline{I}}{\partial x^2}-\frac{\partial^2 \underline{I}}{\partial x^2}\right)+\frac{\partial^2 \underline{I}}{\partial x^2}\right\}=\frac{\partial^2 \tilde{I}\left(x,t\,;\gamma\right)}{\partial x^2}, \qquad (4.13)$$

$$\left\{\beta\left(\frac{\partial^2 \overline{U}}{\partial x^2}-\frac{\partial^2 \underline{U}}{\partial x^2}\right)+\frac{\partial^2 \underline{U}}{\partial x^2}\right\}=\frac{\partial^2 \tilde{U}\left(x,t\,;\gamma\right)}{\partial x^2}, \qquad (4.14)$$
$$\{\beta\left(0.2-0.2\gamma\right)+0.1\gamma+0.71\}=\eta_1,$$

$$\{\beta\left(0.2-0.2\gamma\right)+0.1\gamma+0.9\}=\eta_2,$$
$$\left(\beta\left(0.2-0.2\gamma\right)+0.1\gamma+0.1\right)-a_1=\eta_3. \qquad (4.15)$$

Plugging Eqs. (4.9)– (4.15) into Eq. (4.8), we obtain the following expressions:

$$\begin{cases} \frac{\partial^\alpha \tilde{N}(x,t\,;\gamma,\beta)}{\partial t^\alpha}=\lambda_1\tilde{N}\left(x,t\,;\,\gamma,\beta\right)-r_2\eta_1\,\tilde{N}^2\left(x,t\,;\,\gamma,\beta\right)-c_4\tilde{T}\left(x,t\,;\,\gamma,\beta\right)\tilde{N}\left(x,t\,;\,\gamma,\beta\right) \\ \qquad +a_3e^{-\tilde{U}(x,t\,;\gamma,\beta)}\tilde{N}\left(x,t\,;\,\gamma,\beta\right)+D_N\frac{\partial^2 \tilde{N}(x,t\,;\gamma,\beta)}{\partial x^2},\ 0<\alpha\leq 1, \\ \frac{\partial^\alpha \tilde{T}(x,t\,;\gamma,\beta)}{\partial t^\alpha}=\lambda_2\tilde{T}\left(x,t\,;\,\gamma,\beta\right)-r_1\eta_2\tilde{T}^2\left(x,t\,;\,\gamma,\beta\right)-c_2\tilde{I}\left(x,t\,;\,\gamma,\beta\right)\tilde{T}\left(x,t\,;\,\gamma,\beta\right) \\ \qquad -c_3\tilde{T}\left(x,t\,;\,\gamma,\beta\right)\tilde{N}\left(x,t\,;\,\gamma,\beta\right)+a_2e^{-\tilde{U}(x,t\,;\gamma,\beta)}\tilde{T}\left(x,t\,;\,\gamma,\beta\right) \\ \qquad +D_T\frac{\partial^2 \tilde{T}(x,t\,;\gamma,\beta)}{\partial x^2}, \\ \frac{\partial^\alpha \tilde{I}(x,t\,;\gamma,\beta)}{\partial t^\alpha}=\varepsilon+\frac{\rho\tilde{I}(x,t\,;\gamma,\beta)\,\tilde{T}(x,t\,;\gamma,\beta)}{\mu+\tilde{T}(x,t\,;\gamma,\beta)}-c_1\tilde{I}\left(x,t\,;\,\gamma,\beta\right)\tilde{T}\left(x,t\,;\,\gamma,\beta\right)-\eta_3 \\ \qquad \tilde{I}\left(x,t\,;\,\gamma,\beta\right)+a_1e^{-\tilde{U}(x,t\,;\gamma,\beta)}\tilde{I}\left(x,t\,;\,\gamma,\beta\right)+D_I\frac{\partial^2 \tilde{I}(x,t\,;\gamma,\beta)}{\partial x^2}, \\ \frac{\partial^\alpha \tilde{U}(x,t\,;\gamma,\beta)}{\partial t^\alpha}=1-\eta_2\,\tilde{U}\left(x,t\,;\,\gamma,\beta\right)+D_U\frac{\partial^2 \tilde{U}(x,t\,;\gamma,\beta)}{\partial x^2},\quad -2\leq x\leq 2, \end{cases}$$
$$(4.16)$$

subject to initial conditions Eq. (4.2).

Now, using FRDTM to Eq. (4.16) and initial condition Eq. (4.2), we have the following expressions:

$$
\begin{cases}
\tilde{N}_{k+1}(x;\gamma,\beta) = \frac{(1+\alpha k)}{(1+\alpha k+\alpha)} \\
\quad \left\{
\begin{array}{l}
\lambda_1 \tilde{N}_k(x;\gamma,\beta) - r_2\eta_1 \sum_{i=0}^{k} \tilde{N}_i(x;\gamma,\beta)\,\tilde{N}_{k-i}(x;\gamma,\beta) \\
-c_4 \sum_{i=0}^{k} \tilde{T}_i(x;\gamma,\beta)\,\tilde{N}_{k-i}(x;\gamma,\beta) + a_3 \\
\sum_{i=0}^{k} e^{-\tilde{U}_i(x;\gamma,\beta)}\tilde{N}_{k-i}(x;\gamma,\beta) + D_N \frac{\partial^2 \tilde{N}_k(x;\gamma,\beta)}{\partial x^2}
\end{array}
\right\}, \\
\tilde{T}_{k+1}(x;\gamma,\beta) = \frac{(1+\alpha k)}{(1+\alpha k+\alpha)} \\
\quad \left\{
\begin{array}{l}
\lambda_2 \tilde{T}_k(x;\gamma,\beta) - r_1\eta_2 \sum_{i=0}^{k} \tilde{T}_i(x;\gamma,\beta)\,\tilde{T}_{k-i}(x;\gamma,\beta) \\
-c_2 \sum_{i=0}^{k} \tilde{I}_i(x;\gamma,\beta)\,\tilde{T}_{k-i}(x;\gamma,\beta) - c_3 \\
\sum_{i=0}^{k} \tilde{T}_i(x;\gamma,\beta)\,\tilde{N}_{k-i}(x;\gamma,\beta) + a_2 \\
\sum_{i=0}^{k} e^{-\tilde{U}_i(x;\gamma,\beta)}\tilde{T}_{k-i}(x;\gamma,\beta) + D_T \frac{\partial^2 \tilde{T}_k(x;\gamma,\beta)}{\partial x^2}
\end{array}
\right\}, \\
\tilde{I}_{k+1}(x;\gamma,\beta) = \frac{(1+\alpha k)}{(1+\alpha k+\alpha)} \\
\quad \left\{
\begin{array}{l}
\varepsilon\delta(k) + \rho \sum_{i=0}^{k} \tilde{I}_i(x;\gamma,\beta)\,\tilde{T}_{k-i}(x;\gamma,\beta)\left(\mu + \tilde{T}_i(x;\gamma,\beta)\right)^{-1} \\
-c_1 \sum_{i=0}^{k} \tilde{I}_i(x;\gamma,\beta)\,\tilde{T}_{k-i}(x;\gamma,\beta) - \eta_3 \tilde{I}_k(x;\gamma,\beta) \\
+a_1 \sum_{i=0}^{k} e^{-\tilde{U}_i(x;\gamma,\beta)}\tilde{I}_{k-i}(x;\gamma,\beta) + D_I \frac{\partial^2 \tilde{I}_k(x;\gamma,\beta)}{\partial x^2}
\end{array}
\right\}, \\
\tilde{U}_{k+1}(x;\gamma,\beta) = \frac{(1+\alpha k)}{(1+\alpha k+\alpha)} \\
\quad \left\{\delta(k) - \eta_2 \tilde{U}_k(x;\gamma,\beta) + D_U \frac{\partial^2 \tilde{U}_k(x;\gamma,\beta)}{\partial x^2}\right\}, \quad \text{for} \quad k = 0,1,2,\dots
\end{cases}
\tag{4.17}
$$

with transformed initial conditions:

$$
\begin{aligned}
\tilde{N}_0(x;\gamma,\beta) &= 0.2e^{-2x^2}, \quad \tilde{T}_0(x;\gamma,\beta) = 1 - 0.75\,\sec h(x), \\
\tilde{I}_0(x;\gamma,\beta) &= 0.375 - 0.235\,\sec h^2(x), \quad \tilde{U}_0(x;\gamma,\beta) = \sec h(x),
\end{aligned}
\tag{4.18}
$$

where

$$
\delta(k) = \begin{cases} 1 & k = 0, \\ 0 & k \neq 0. \end{cases}
\tag{4.19}
$$

Substituting transformed initial conditions Eq. (4.18) into Eq. (4.17) and using the values of the parameters given in Table 4.1, the following expressions for $k = 0,1,2,\dots$ are obtained successively:

$$\tilde{N}_1(x;\gamma,\beta) = \frac{1}{\Gamma(1+\alpha)}$$
$$\left\{ \begin{array}{l} 0.1792e^{-2x^2} - 0.04\eta_1\left(e^{-2x^2}\right)^2 - 0.2\left(1 - 0.75\sec h(x)\right)e^{-2x^2} \\ +0.02e^{-\sec h(x)}e^{-2x^2} + 0.0032x^2e^{-2x^2} \end{array} \right\}, \qquad (4.20)$$

$$\tilde{T}_1(x;\gamma,\beta) = \frac{1}{\Gamma(1+\alpha)}$$
$$\left\{ \begin{array}{l} 0.8 - 0.6\sec h(x) - 1.1\,\eta_2\left(1 - 0.75\sec h(x)\right)^2 - 0.55 \\ \left(0.375 - 0.235\sec h(x)^2\right)\left(1 - 0.75\sec h(x)\right) - 0.18 \\ \left(1 - 0.75\sec h(x)\right)e^{-2x^2} + 0.3e^{-\sec h(x)}\left(1 - 0.75\sec h(x)\right) - 0.00075 \\ \sec h(x)\tanh(x)^2 + 0.00075\sec h(x)\left(1 - \tanh(x)^2\right) \end{array} \right\}, \qquad (4.21)$$

$$\tilde{I}_1(x;\gamma,\beta) = \frac{1}{\Gamma(1+\alpha)}$$
$$\left\{ \begin{array}{l} 0.33 - \frac{0.2\left(0.375 - 0.235\sec h(x)^2\right)\left(1 - 0.75\sec h(x)\right)}{\left(1.3 - 0.75\sec h(x)\right)} - \\ \left(0.375 - 0.235\sec h(x)^2\right)\left(1 - 0.75\sec h(x)\right) - \eta_3\left(\begin{array}{l} 0.375 - \\ 0.235\sec h(x)^2 \end{array} \right) \\ +0.2e^{-\sec h(x)}\left(0.375 - 0.235\sec h(x)^2\right) - 0.00094\sec h(x)^2 \\ \tanh(x)^2 + 0.000470\sec h(x)^2\left(1 - \tanh(x)^2\right) \end{array} \right\}, \qquad (4.22)$$

$$\tilde{U}_1(x;\gamma,\beta) = \frac{1}{\Gamma(1+\alpha)}$$
$$\left\{ \begin{array}{l} -\eta_2\sec h(x) + 1 + 0.001\sec h(x)\tanh(x)^2 - 0.001\sec h(x) \\ \left(1 - \tanh(x)^2\right) \end{array} \right\}. \qquad (4.23)$$

Proceeding likewise, one may obtain all the values of $\{N_k\}_{k=0}^{\infty}$, $\{T_k\}_{k=0}^{\infty}$, $\{I_k\}_{k=0}^{\infty}$ and $\{U_k\}_{k=0}^{\infty}$. Applying inverse differential transform to $\{N_k\}_{k=0}^{\infty}$, $\{T_k\}_{k=0}^{\infty}$, $\{I_k\}_{k=0}^{\infty}$, $\{U_k\}_{k=0}^{\infty}$, and using the values of η_1, η_2, and η_3, we have the following nth order approximate solutions:

$$\left\{ \begin{array}{l} \tilde{N}_n(x,t;\gamma,\beta) = \sum_{k=0}^{n}\tilde{N}_k(x;\gamma,\beta)\,t^{\alpha k}, \\ \tilde{T}_n(x,t;\gamma,\beta) = \sum_{k=0}^{n}\tilde{T}_k(x;\gamma,\beta)\,t^{\alpha k}, \\ \tilde{I}_n(x,t;\gamma,\beta) = \sum_{k=0}^{n}\tilde{I}_k(x;\gamma,\beta)\,t^{\alpha k}, \\ \tilde{U}_n(x,t;\gamma,\beta) = \sum_{k=0}^{n}\tilde{U}_k(x;\gamma,\beta)\,t^{\alpha k}. \end{array} \right. \qquad (4.24)$$

One may write the exact solution of this model as

$$
\begin{cases}
\tilde{N}(x,t;\gamma,\beta) = \lim_{n\to\infty} \tilde{N}_n(x,t;\gamma,\beta), & \tilde{T}(x,t;\gamma,\beta) = \lim_{n\to\infty} \tilde{T}_n(x,t;\gamma,\beta), \\
\tilde{I}(x,t;\gamma,\beta) = \lim_{n\to\infty} \tilde{I}_n(x,t;\gamma,\beta), & \tilde{U}(x,t;\gamma,\beta) = \lim_{n\to\infty} \tilde{U}_n(x,t;\gamma,\beta).
\end{cases} \tag{4.25}
$$

The lower- and upper-bound solutions of the model can respectively be obtained by substituting $\beta = 0$ and $\beta = 1$ in Eq. (4.25). Mathematically, we may respectively write the following equations:

$$
\begin{cases}
\tilde{N}(x,t;\gamma,0) = \underline{N}(x,t;\gamma), & \tilde{T}(x,t;\gamma,0) = \underline{T}(x,t;\gamma), \\
\qquad \tilde{I}(x,t;\gamma,0) = \underline{I}(x,t;\gamma), & \tilde{U}(x,t;\gamma,0) = \underline{U}(x,t;\gamma), \\
\text{and} \\
\tilde{N}(x,t;\gamma,1) = \overline{N}(x,t;\gamma), & \tilde{T}(x,t;\gamma,1) = \overline{T}(x,t;\gamma), \\
\qquad \tilde{I}(x,t;\gamma,1) = \overline{I}(x,t;\gamma), & \tilde{U}(x,t;\gamma,1) = \overline{U}(x,t;\gamma).
\end{cases} \tag{4.26}
$$

4.4   RESULTS AND DISCUSSION

Various numerical computations have been carried out by taking different values of parameters involved in the governing model. The values of the parameters are taken as given in Table 4.1, which are stated in the references [5, 6]. Here, all the numerical results and plots are included by considering the second-order ($n = 2$) solutions. As mentioned earlier, the carrying capacity b_1, b_2 and the death rate are assumed to be TFN. Accordingly, the triangular fuzzy solutions of the titled model are depicted in Fig. 4.1 by changing the values of $x \in [-2, 2]$ at $\alpha = 1$ and $t = 0.25$. From Fig. 4.2, one may observe that the crisp result ($\gamma = 1$) is the central line, and all the interval solutions are spread on both sides of the crisp results. It is worth mentioning that the present results at $\gamma = 1$ exactly match the second-order approximate solution of Veeresha et al. [5]. Figure 4.3 illustrates the solution plots of the titled model at different values of fractional order (α) ($= 0.75, 0.9$). It is noticed from Fig. 4.3 that the considered model significantly depends on the time-fractional order derivatives, which helps to analyze the biological behavior of the tumor and immune cells with chemotherapeutic drugs. From Figs. 4.3a and 4.3b, one may see that tumor cells and immune cells are least at $x = 0$ and increasing from the center to both sides with increasing the values of x. But, for normal cells and chemotherapeutic drugs, it is peak at $x = 0$ and decreases with increasing the values of x. Table 4.2 includes the lower- and upper-bound fuzzy solutions of the model at different values of x, t, γ, and $\alpha = 1$. From Table 4.2, it is clear that the lower-bound and the upper-bound are equal at $\gamma = 1$ and these values are the same as the solution of Veeresha et al [5]. In a similar fashion, one may consider other involved parameters as well as initial conditions as different fuzzy numbers. The methodology may easily then be extended accordingly.

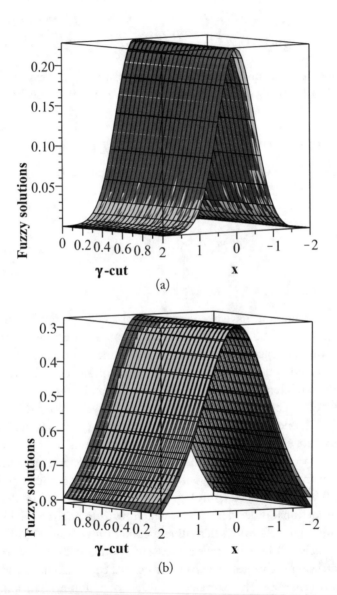

Figure 4.1: Lower- and upper-bound fuzzy solutions of Eq. (4.16): (a) normal cells (N) and (b) tumor cells (T), $\alpha = 1$, $t = 0.25$, and $x \in [-2, 2]$. (*Continues.*)

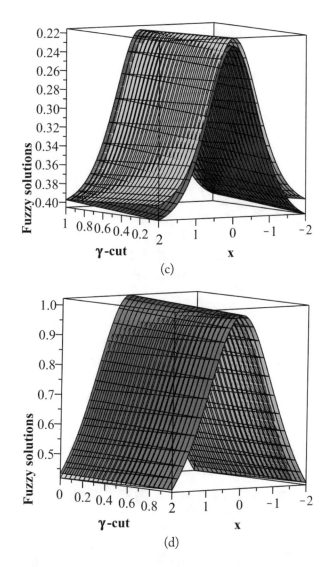

Figure 4.1: (*Continued.*) Lower- and upper-bound fuzzy solutions of Eq. (4.16): (c) immune cells (I) and (d) chemotherapeutic drug (U) at $\alpha = 1$, $t = 0.25$, and $x \in [-2, 2]$.

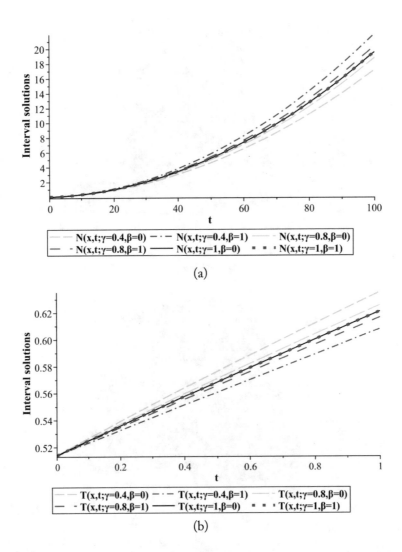

Figure 4.2: Interval solutions of Eq. (4.16): (a) normal cells (N) and (b) tumor cells (T) when $\alpha = 1$, $x = 1$ at different values of γ-cut. (*Continues.*)

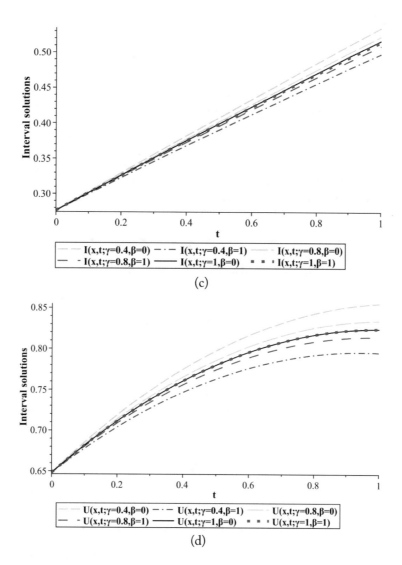

Figure 4.2: (*Continued.*) Interval solutions of Eq. (4.16): (c) immune cells (I) and (d) chemotherapeutic drug (U) when $\alpha = 1$, $x = 1$ at different values of γ-cut.

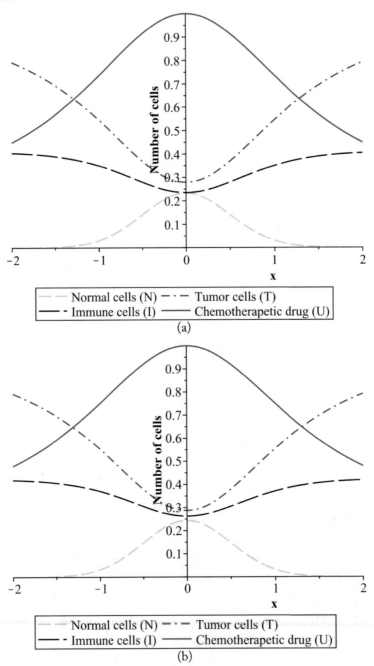

Figure 4.3: Solutions plots of Eq. (4.16) at (a) $\alpha = 0.9$ and (b) $\alpha = 0.75$ when $t = 0.25$, $\gamma = 1$, and $-2 \leq x \leq 2$.

Table 4.2: The fuzzy and crisp solution of the model at $\alpha = 1$

$t \rightarrow$ $x \downarrow$		0.2	0.4	0.8	1.0
0.2, $\gamma = 0$	$[\underline{N}, \overline{N}]$	[0.2052, 0.2038]	[0.2273, 0.2244]	[0.2760, 0.2695]	[0.3025, 0.2941]
	$[\underline{T}, \overline{T}]$	[0.2864, 0.2835]	[0.3087, 0.3032]	[0.3549, 0.3453]	[0.3789, 0.3676]
	$[\underline{I}, \overline{I}]$	[0.2159, 0.2089]	[0.2838, 0.2678]	[0.4231, 0.3829]	[0.4944, 0.4391]
	$[\underline{U}, \overline{U}]$	[1.0016, 0.9662]	[1.0187, 0.9556]	[1.0403, 0.9451]	[1.0449, 0.9451]
0.4, $\gamma = 0.4$	$[\underline{N}, \overline{N}]$	[0.1609, 0.1604]	[0.1778, 0.1767]	[0.2147, 0.2126]	[0.2348, 0.2322]
	$[\underline{T}, \overline{T}]$	[0.3296, 0.3272]	[0.3535, 0.3489]	[0.4028, 0.3944]	[0.4283, 0.4182]
	$[\underline{I}, \overline{I}]$	[0.2378, 0.2331]	[0.3022, 0.2916]	[0.4322, 0.4065]	[0.4978, 0.4628]
	$[\underline{U}, \overline{U}]$	[0.9485, 0.9283]	[0.9672, 0.9309]	[0.9900, 0.9339]	[0.9941, 0.9934]
0.8, $\gamma = 0.8$	$[\underline{N}, \overline{N}]$	[0.0610, 0.0609]	[0.0667, 0.0667]	[0.0792, 0.0792]	[0.0860, 0.0861]
	$[\underline{T}, \overline{T}]$	[0.4642, 0.4625]	[0.4891, 0.4859]	[0.5389, 0.5329]	[0.5637, 0.5565]
	$[\underline{I}, \overline{I}]$	[0.2992, 0.2971]	[0.3541, 0.3496]	[0.4619, 0.4517]	[0.5148, 0.5013]
	$[\underline{U}, \overline{U}]$	[0.7958, 0.7903]	[0.8336, 0.8232]	[0.8776, 0.8600]	[0.8839, 0.8640]
1.0, $\gamma = 1.0$	$[\underline{N}, \overline{N}]$	[0.0294, 0.0294]	[0.0319, 0.0319]	[0.0375, 0.0375]	[0.0404, 0.0404]
	$[\underline{T}, \overline{T}]$	[0.5359, 0.5359]	[0.5576, 0.5576]	[0.6001, 0.6001]	[0.6209, 0.6209]
	$[\underline{I}, \overline{I}]$	[0.3261, 0.3261]	[0.3751, 0.3751]	[0.4704, 0.4704]	[0.5167, 0.5167]
	$[\underline{U}, \overline{U}]$	[0.7114, 0.7114]	[0.7607, 0.7607]	[0.8170, 0.8170]	[0.8240, 0.8240]
Veeresha et al.,	[N]	0.0294	0.0319	0.0375	0.0404
(Crisp) [5]	[T]	0.5359	0.5576	0.6001	0.6209
	[I]	0.3262	0.3756	0.4722	0.5195
	[U]	0.7114	0.7607	0.8170	0.8240

4.5   REFERENCES

[1] V. A. Kuznetsov and G. D. Knott. Modeling tumor regrowth and immunotherapy, *Mathematical and Computer Modelling*, 33:1275–1287, 2001. DOI: 10.1016/s0895-7177(00)00314-9. 37

[2] J. Adam and J. Panetta. A simple mathematical model and alternative paradigm for certain chemotherapeutic regimens. *Mathematical and Computer Modelling*, 22(8):49–60, 1995. DOI: 10.1016/0895-7177(95)00154-t.

[3] R. D. Boer, P. Hogeweg, H. Dullens, R. A. D. Weger, W. D. Otter. Macrophage T lymphocyte interactions in the anti-tumor immune response: A mathematical model. *Journal of Immunology*, 134(4):2748–2758, 1985.

[4] D. Kirschner and J. C. Panetta. Modeling immunotherapy of the tumor-immune interaction. *Journal of Mathematical Biology*, 37(3):235–252, 1998. DOI: 10.1007/s002850050127. 37

[5] P. Veeresha, D. G. Prakasha, and H. M. Baskonus. New numerical surfaces to the mathematical model of cancer chemotherapy effect in Caputo fractional derivatives. *Chaos*, 29(013119):1–14, 2019. DOI: 10.1063/1.5074099. 37, 38, 45

[6] V. F. Morales-Delgado, J. F. Gómez-Aguilar, K. Saad, and R. F. E Jiménez. Application of the Caputo–Fabrizio and Atangana–Baleanu fractional derivatives to mathematical model of cancer chemotherapy effect. *Mathematical Methods in Applied Sciences*, 42:1–27, 2019. DOI: 10.1002/mma.5421. 38, 45

[7] Y. Louzoun, C. Xue, G. B. Lesinski, and A. Friedman. A mathematical model for pancreatic cancer growth and treatments. *Journal of Theoretical Biology*, 351:74–82, 2014. DOI: 10.1016/j.jtbi.2014.02.028. 37

[8] S. L. Weekes, B. Barker, S. Bober, et al. A multicompartment mathematical model of cancer stem cell-driven tumor growth dynamics. *Bulletin of Mathematical Biophysics*, 76(7):1762–1782, 2014. DOI: 10.1007/s11538-014-9976-0. 37

[9] M. Usman, G. Flora, C. Yakopcic, and M. Imran. A computational study and stability analysis of a mathematical model for in vitro inhibition of cancer cell mutation. *International Journal of Applied and Computational Mathematics*, 3(3):1861–1878, 2017. DOI: 10.1007/s40819-016-0201-8. 37

[10] K. Abernathy, Z. Abernathy, A. Baxter, and M. Stevens. Global dynamics of a breast cancer competition model. *Differential Equations and Dynamical Systems*, 1:1–15, 2017. DOI: 10.1007/s12591-017-0346-x. 37

[11] A. Mondol, R. Gupta, S. Das, and T. Dutta. An insight into Newton's cooling law using fractional calculus. *Journal of Applied Physics*, 123(6):3–10, 2018. DOI: 10.1063/1.4998236. 37

[12] R. M. Jena, S. Chakraverty, and S. K. Jena. Dynamic response analysis of fractionally damped beams subjected to external loads using Homotopy Analysis Method (HAM). *Journal of Applied and Computational Mechanics*, 5(2):355–366, 2019. DOI: 10.22055/jacm.2019.27592.1419. 37

[13] R. M. Jena and S. Chakraverty. Residual power series method for solving time-fractional model of vibration equation of large membranes. *Journal of Applied and Computational Mechanics*, 5(4):603–615, 2019. DOI: 10.22055/jacm.2018.26668.1347.

[14] R. M. Jena, S. Chakraverty, and D. Baleanu. On new solutions of time-fractional wave equations arising in Shallow water wave propagation. *Mathematics*, 7:722, 2019. DOI: 10.3390/math7080722.

[15] R. M. Jena and S. Chakraverty. Solving time-fractional Navier–Stokes equations using homotopy perturbation Elzaki transform. *SN Applied Sciences*, 1(1):16, 2019. DOI: 10.1007/s42452-018-0016-9.

[16] R. M. Jena and S. Chakraverty. A new iterative method based solution for fractional Black–Scholes option pricing equations (BSOPE). *SN Applied Sciences*, 1(1):95, 2019. DOI: 10.1007/s42452-018-0106-8. 37

[17] H. Namazi, V. V. Kulish, and A. Wong. Mathematical modelling and prediction of the effect of chemotherapy on cancer cells. *Scientific Reports*, 5:1–8, 2015. DOI: 10.1038/srep13583. 38

[18] S. Fahmy, A. M. El-Geziry, E. Mohamed, A. M. Abdel-Aty, and A. G. Radwan. Fractional-order mathematical model for chronic myeloid leukaemia. *17th European Conference on Circuit Theory and Design (ECCTD)*. IEEE, 2017. DOI: 10.1109/ecctd.2017.8093247.

[19] F. Bozkurt, T. Abdeljawad, and M. A. Hajji. Stability analysis of a fractional-order differential equation model of a brain tumor growth depending on the density. *Applied and Computational Mathematics*, 14(1):50–62, 2015. 38

[20] F. Ansarizadeh, M. Singh, and D. Richards. Modelling of tumor cells regression in response to chemotherapeutic treatment. *Applied Mathematical Modelling*, 48:96–112, 2017. DOI: 10.1016/j.apm.2017.03.045. 38

[21] J. C. Panetta. A mathematical model of periodically pulsed chemotherapy tumor recurrence and metastasis in a competitive environment. *Bulletin of Mathematical Biology*, 58(3):425–447, 1996. DOI: 10.1016/0092-8240(95)00346-0. 38

[22] V. A. Kuznetsov, I. A. Makalkin, M. A. Taylor, and A. S. Perelson. Nonlinear dynamics of immunogenic tumors: Parameter estimation and global bifurcation analysis. *Bulletin of Mathematical Biology*, 56(2):295–321, 1994. DOI: 10.1016/s0092-8240(05)80260-5.

[23] R. M. Jena, S. Chakraverty, and D. Baleanu. On the solution of imprecisely defined nonlinear time-fractional dynamical model of marriage. *Mathematics*, 7:689–704, 2019. DOI: 10.3390/math7080689.

[24] R. M. Jena and S. Chakraverty. Analytical solution of Bagley–Torvik equations using Sumudu transformation method. *SN Applied Sciences*, 1(3):246, 2019. DOI: 10.1007/s42452-019-0259-0.

[25] S. Chakraverty, S. Tapaswini, and D. Behera. *Fuzzy Arbitrary Order System: Fuzzy Fractional Differential Equations and Applications.* John Wiley & Sons, 2016. DOI: 10.1002/9781119004233.

[26] S. Chakraverty, S. Tapaswini, and D. Behera. *Fuzzy Differential Equations and Applications for Engineers and Scientists.* Taylor & Francis Group, CRC Press, Boca Raton, FL, 2016. DOI: 10.1201/9781315372853.

[27] S. Chakraverty, D. M. Sahoo, and N. R. Mahato. *Concepts of Soft Computing: Fuzzy and ANN with Programming.* Springer, Singapore, 2019. DOI: 10.1007/978-981-13-7430-2.

CHAPTER 5

Fuzzy Time-Fractional Smoking Epidemic Model

5.1 INTRODUCTION

In 1766, Swiss mathematician and physicist Bernoulli [1] developed the idea of mathematical modeling for the transmission of diseases, which was the start of modern epidemiology. Further, Ross [2] provided the modeling of infectious disease at the beginning of the 20th century and described the behaviors of epidemic models by using the law of mass action. Epidemic models have been widely utilized to study epidemiological processes, which include transmission of contagious diseases. These types of models have also been implemented to study the spread of social habits such as alcohol consumption [3], obesity epidemics [4], cocaine consumption [5], smoking [6], and many more. Among these models, the Smoking Epidemic Model (SEM) has been one of the most challenging for the last few decades. The World Health Organization (WHO) proclaimed that smoking causes 250 million child and adolescent deaths and predicted that more than 10 million people would perish due to smoking-related diseases every year by 2030 [7]. The main effects of short-term smoking are high blood pressure, stained teeth, bad breath, and coughing. In recent years, lung cancer, throat cancer, mouth cancer, stomach ulcers, heart disease, and gum disease have been the principal threats from long-term smoking. Therefore, smoking is treated as a leading global public-health problem. Smoking can also spread similarly to the spread of many infectious diseases through social contact. Thus, mathematical modeling has been widely used to investigate the behavior of smoking.

In this chapter, the titled problem has been modeled using the Caputo fractional derivative. The main aim of this investigation is to study the dynamics of time-fractional SEM with uncertainty. Here, we have considered the parameters involved in the model as triangular fuzzy numbers.

5.2 MATHEMATICAL MODEL

In the present study, we have considered the SEM in time-fractional order derivatives. The SEM is divided into five subgroups viz. potential smoker $P(t)$, the occasional smoker $O(t)$, smoker $S(t)$, temporarily quit smoker $Q(t)$, and permanently quit smoker $L(t)$. Let us consider the total number of the population at time t is $N(t)$. So, we can write this mathematically $N(t) = P(t) + O(t) + S(t) + Q(t) + L(t)$. The proposed time-fractional SEM defined in

Caputo sense is written as [9, 10]:

$$\begin{cases} D_t^{\mu} P\,(t) = \lambda - bP\,(t)\,S(t) - \eta P\,(t)\,, \\ D_t^{\mu} O\,(t) = bP\,(t)\,S(t) - \alpha_1 O\,(t) - \eta O\,(t)\,, \\ D_t^{\mu} S\,(t) = \alpha_1 O\,(t) + \alpha_2 S\,(t)\,Q\,(t) - (\eta + r)\,S(t)\,, \qquad 0 < \mu \leq 1 \qquad (5.1) \\ D_t^{\mu} Q\,(t) = -\alpha_2 S\,(t)\,Q\,(t) - \eta Q\,(t) + r\,(1 - \delta)\,S(t)\,, \\ D_t^{\mu} L\,(t) = \delta\,r S(t) - \eta L\,(t)\,, \end{cases}$$

subject to initial conditions [9]:

$$P\,(0) = 40, \quad O\,(0) = 10, \quad S\,(0) = 20, \quad Q\,(0) = 10, \quad \text{and} \quad L\,(0) = 5, \qquad (5.2)$$

where λ denotes the recruitment rate in P, b indicates the effective contact rate between $S\,(t)$ and $P\,(t)$, η symbolizes the natural death rate, r stands for rate of quitting smoking, δ represents the remaining fraction of smoking who permanently quit smoking, α_1 is the rate at which occasional smokers become regular smokers and α_2 indicates the contact rate between smoker and temporary quitters who revert to smoking.

5.3   EQUILIBRIUM POINT AND STABILITY

For the equilibrium point, we consider Eq. (5.1) as follows:

$$D_t^{\mu} P\,(t) = D_t^{\mu} O\,(t) = D_t^{\mu} S\,(t) = D_t^{\mu} Q\,(t) = D_t^{\mu} L\,(t) = 0. \qquad (5.3)$$

In Eq. (5.1), we find disease-free equilibria as [9]

$$E_0 = (P, 0, S, 0, 0)\,,$$

and the endemic equilibria of Eq. (5.1) is

$$E^* = \left(P^*, P^*, S^*, Q^*, L^*\right),$$

where

$$P^* = \frac{\lambda}{bS\,(t) + \mu\,e}, \quad O^* = \frac{b\lambda S\,(t)}{(bS\,(t) + \mu)\,(\alpha_1 + \mu)}, \\ Q^* = \frac{r\,(1 - \delta)\,S\,(t)}{\alpha_2 S\,(t) + \mu}, \quad L^* = \frac{\delta\,r S\,(t)}{\mu}. \qquad (5.4)$$

Theorem 5.1   *The disease-free equilibrium E^* is locally asymptotically stable if $R_0 < 1$ and is unstable if $R_0 > 1$, where R_0 is a reproductive number [10, 11] and defined as*

$$R_0 = \frac{\alpha_1 b\mu\,P\,(t)}{(\alpha_1 + \mu)\,(\mu + r)\,(\mu + bS\,(t))}. \qquad (5.5)$$

5.4   MATHEMATICAL MODEL WITH FUZZY PARAMETERS

In this section, the fuzzy fractional epidemic model is first converted to interval-based fuzzy differential equations using single parametric form. Then, using the double parametric form, the interval-based fuzzy differential equations are reduced to the parametric form of differential equations. Next, by applying FRDTM to the parametric form of differential equations, we may obtain the fuzzy solution of the model. In the present investigation, we have considered η, α_1, and α_2 as TFN. Let us consider the following fuzzy fractional SEM as

$$
\begin{cases}
D_t^{\mu} \tilde{P}(t) = \lambda - b \tilde{P}(t) \tilde{S}(t) - (\eta - 0.01, \eta, \eta + 0.01) \, \tilde{P}(t), \\
D_t^{\mu} \tilde{O}(t) = b \tilde{P}(t) \tilde{S}(t) - (\alpha_1 - 0.001, \alpha_1, \alpha_1 + 0.001) \, \tilde{O}(t) \\
\qquad\quad - (\eta - 0.01, \eta, \eta + 0.01) \, \tilde{O}(t), \\
D_t^{\mu} \tilde{S}(t) = (\alpha_1 - 0.001, \alpha_1, \alpha_1 + 0.001) \, \tilde{O}(t) + (\alpha_2 - 0.0001, \alpha_2, \alpha_2 + 0.0001) \\
\qquad\quad \tilde{S}(t) \tilde{Q}(t) - ((\eta - 0.01, \eta, \eta + 0.01) + r) \, \tilde{S}(t), \\
D_t^{\mu} \tilde{Q}(t) = -(\alpha_2 - 0.0001, \alpha_2, \alpha_2 + 0.0001) \, \tilde{S}(t) \tilde{Q}(t) \\
\qquad\quad - (\eta - 0.01, \eta, \eta + 0.01) \, \tilde{Q}(t) + r (1 - \delta) \, \tilde{S}(t), \\
D_t^{\mu} \tilde{L}(t) = \delta \, r \tilde{S}(t) - (\eta - 0.01, \eta, \eta + 0.01) \, \tilde{L}(t)
\end{cases}
\tag{5.6}
$$

subject to initial conditions Eq. (5.2).

Using γ-cut, Eq. (5.6) is reduced to the following expressions:

$$
\begin{cases}
\left[D_t^\mu \underline{P}(t;\gamma), D_t^\mu \overline{P}(t;\gamma)\right] = \lambda - b\left[\underline{P}(t;\gamma), \overline{P}(t;\gamma)\right]\left[\underline{S}(t;\gamma), \overline{S}(t;\gamma)\right] \\[2mm]
\quad - \begin{bmatrix} 0.01\gamma + \eta - 0.01, \\ -0.01\gamma + \eta + 0.01 \end{bmatrix}\left[\underline{P}(t;\gamma), \overline{P}(t;\gamma)\right], \\[4mm]
\left[D_t^\mu \underline{O}(t;\gamma), D_t^\mu \overline{O}(t;\gamma)\right] = b\left[\underline{P}(t;\gamma), \overline{P}(t;\gamma)\right]\left[\underline{S}(t;\gamma), \overline{S}(t;\gamma)\right] \\[2mm]
\quad - \begin{bmatrix} 0.001\gamma + \alpha_1 - 0.001, \\ -0.001\gamma + \alpha_1 + 0.001 \end{bmatrix} \\[2mm]
\quad \left[\underline{O}(t;\gamma), \overline{O}(t;\gamma)\right] - [0.01\gamma + \eta - 0.01, -0.01\gamma + \eta + 0.01]\left[\underline{O}(t;\gamma), \overline{O}(t;\gamma)\right], \\[6mm]

\left[D_t^\mu \underline{S}(t;\gamma), D_t^\mu \overline{S}(t;\gamma)\right] = [0.001\gamma + \alpha_1 - 0.001, -0.001\gamma + \alpha_1 + 0.001] \\[2mm]
\quad \left[\underline{O}(t;\gamma), \overline{O}(t;\gamma)\right] + [0.0001\gamma + \alpha_2 - 0.0001, -0.0001\gamma + \alpha_2 + 0.0001] \\[2mm]
\quad \left[\underline{S}(t;\gamma), \overline{S}(t;\gamma)\right]\left[\underline{Q}(t;\gamma), \overline{Q}(t;\gamma)\right] \\[2mm]
\quad - ([0.01\gamma + \eta - 0.01, -0.01\gamma + \eta + 0.01] + r)\left[\underline{S}(t;\gamma), \overline{S}(t;\gamma)\right], \\[6mm]

\left[D_t^\mu \underline{Q}(t;\gamma), D_t^\mu \overline{Q}(t;\gamma)\right] = -[0.0001\gamma + \alpha_2 - 0.0001, -0.0001\gamma + \alpha_2 + 0.0001] \\[2mm]
\quad \left[\underline{S}(t;\gamma), \overline{S}(t;\gamma)\right]\left[\underline{Q}(t;\gamma), \overline{Q}(t;\gamma)\right] - [0.01\gamma + \eta - 0.01, -0.01\gamma + \eta + 0.01] \\[2mm]
\quad \left[\underline{Q}(t;\gamma), \overline{Q}(t;\gamma)\right] + r(1-\delta)\left[\underline{S}(t;\gamma), \overline{S}(t;\gamma)\right], \\[6mm]

\left[D_t^\mu \underline{L}(t;\gamma), D_t^\mu \overline{L}(t;\gamma)\right] = \delta r\left[\underline{S}(t;\gamma), \overline{S}(t;\gamma)\right] \\[2mm]
\quad - [0.01\gamma + \eta - 0.01, -0.01\gamma + \eta + 0.01]\left[\underline{L}(t;\gamma), \overline{L}(t;\gamma)\right],
\end{cases}
$$

$$\tag{5.7}$$

where γ is a parameter and $\gamma \in [0, 1]$.

Next, by using the double parametric form, the above Eq. (5.7) may be expressed as

$$
\begin{cases}
\beta \left(D_t^\mu \overline{P}\left(t;\gamma\right) - D_t^\mu \underline{P}\left(t;\gamma\right)\right) + D_t^\mu \underline{P}\left(t;\gamma\right) \\[4pt]
= \lambda - b\left[\beta \left(\overline{P}\left(t;\gamma\right) - \underline{P}\left(t;\gamma\right)\right) + \underline{P}\left(t;\gamma\right)\right] \\[4pt]
\left[\beta \left(\overline{S}\left(t;\gamma\right) - \underline{S}\left(t;\gamma\right)\right) + \underline{S}\left(t;\gamma\right)\right] - \left[\beta\left(-0.02\gamma + 0.02\right) + 0.01\gamma + \eta - 0.01\right] \\[4pt]
\left[\beta \left(\overline{P}\left(t;\gamma\right) - \underline{P}\left(t;\gamma\right)\right) + \underline{P}\left(t;\gamma\right)\right], \\[10pt]
\beta \left(D_t^\mu \overline{O}\left(t;\gamma\right) - D_t^\mu \underline{O}\left(t;\gamma\right)\right) + D_t^\mu \underline{O}\left(t;\gamma\right) \\[4pt]
= b\left[\beta \left(\overline{P}\left(t;\gamma\right) - \underline{P}\left(t;\gamma\right)\right) + \underline{P}\left(t;\gamma\right)\right] \\[4pt]
\left[\beta \left(\overline{S}\left(t;\gamma\right) - \underline{S}\left(t;\gamma\right)\right) + \underline{S}\left(t;\gamma\right)\right] \\[4pt]
- \left[\beta\left(-0.002\gamma + 0.002\right) + 0.001\gamma + \alpha_1 - 0.001\right] \\[4pt]
\left[\beta \left(\overline{O}\left(t;\gamma\right) - \underline{O}\left(t;\gamma\right)\right) + \underline{O}\left(t;\gamma\right)\right] - \left[\beta\left(-0.02\gamma + 0.02\right) + 0.01\gamma + \eta - 0.01\right] \\[4pt]
\left[\beta \left(\overline{O}\left(t;\gamma\right) - \underline{O}\left(t;\gamma\right)\right) + \underline{O}\left(t;\gamma\right)\right], \\[10pt]
\beta \left(D_t^\mu \overline{S}\left(t;\gamma\right) - D_t^\mu \underline{S}\left(t;\gamma\right)\right) + D_t^\mu \underline{S}\left(t;\gamma\right) \\[4pt]
= \left[\beta\left(-0.002\gamma + 0.002\right) + 0.001\gamma + \alpha_1 - 0.001\right] \\[4pt]
\left[\beta \left(\overline{O}\left(t;\gamma\right) - \underline{O}\left(t;\gamma\right)\right) + \underline{O}\left(t;\gamma\right)\right] \\[4pt]
+ \left[\beta\left(-0.0002\gamma + 0.0002\right) + 0.0001\gamma + \alpha_2 - 0.0001\right] \\[4pt]
\left[\beta \left(\overline{S}\left(t;\gamma\right) - \underline{S}\left(t;\gamma\right)\right) + \underline{S}\left(t;\gamma\right)\right]\left[\beta \left(\overline{Q}\left(t;\gamma\right) - \underline{Q}\left(t;\gamma\right)\right) + \underline{Q}\left(t;\gamma\right)\right] \\[4pt]
- \left(\begin{bmatrix} \beta\left(-0.02\gamma + 0.02\right) + \\ 0.01\gamma + \eta - 0.01 \end{bmatrix} + r\right)\left[\beta \left(\overline{S}\left(t;\gamma\right) - \underline{S}\left(t;\gamma\right)\right) + \underline{S}\left(t;\gamma\right)\right], \\[14pt]
\beta \left(D_t^\mu \overline{Q}\left(t;\gamma\right) - D_t^\mu \underline{Q}\left(t;\gamma\right)\right) + D_t^\mu \underline{Q}\left(t;\gamma\right) \\[4pt]
= -\begin{bmatrix} \beta\left(-0.0002\gamma + 0.0002\right) + 0.0001\gamma \\ + \alpha_2 - 0.0001 \end{bmatrix} \\[4pt]
\left[\underline{S}\left(t;\gamma\right), \overline{S}\left(t;\gamma\right)\right]\left[\beta \left(\overline{Q}\left(t;\gamma\right) - \underline{Q}\left(t;\gamma\right)\right) + \underline{Q}\left(t;\gamma\right)\right] \\[4pt]
- \begin{bmatrix} \beta\left(-0.02\gamma + 0.02\right) + 0.01\gamma + \\ \eta - 0.01 \end{bmatrix} \\[4pt]
\left[\beta \left(\overline{Q}\left(t;\gamma\right) - \underline{Q}\left(t;\gamma\right)\right) + \underline{Q}\left(t;\gamma\right)\right] + r\left(1 - \delta\right)\left[\beta \left(\overline{S}\left(t;\gamma\right) - \underline{S}\left(t;\gamma\right)\right) + \underline{S}\left(t;\gamma\right)\right], \\[10pt]
\beta \left(D_t^\mu \overline{L}\left(t;\gamma\right) - D_t^\mu \underline{L}\left(t;\gamma\right)\right) + D_t^\mu \underline{L}\left(t;\gamma\right) = \delta r\left[\beta \left(\overline{S}\left(t;\gamma\right) - \underline{S}\left(t;\gamma\right)\right) + \underline{S}\left(t;\gamma\right)\right] \\[4pt]
- \left[\beta\left(-0.02\gamma + 0.02\right) + 0.01\gamma + \eta - 0.01\right]\left[\beta \left(\overline{L}\left(t;\gamma\right) - \underline{L}\left(t;\gamma\right)\right) + \underline{L}\left(t;\gamma\right)\right],
\end{cases}
$$

$$(5.8)$$

subject to initial condition Eq. (5.2).

Let us symbolize

$$\beta \left(D_t^\mu \overline{P}\,(t\,;\gamma) - D_t^\mu \underline{P}\,(t\,;\gamma) \right) + D_t^\mu \underline{P}\,(t\,;\gamma) = D_t^\mu \tilde{P}\,(t\,;\gamma,\beta),$$

$$\beta \left(D_t^\mu \overline{O}\,(t\,;\gamma) - D_t^\mu \underline{O}\,(t\,;\gamma) \right) + D_t^\mu \underline{O}\,(t\,;\gamma) = D_t^\mu \tilde{O}\,(t\,;\gamma,\beta),$$

$$\beta \left(D_t^\mu \overline{S}\,(t\,;\gamma) - D_t^\mu \underline{S}\,(t\,;\gamma) \right) + D_t^\mu \underline{S}\,(t\,;\gamma) = D_t^\mu \tilde{S}\,(t\,;\gamma,\beta),$$

$$\beta \left(D_t^\mu \overline{Q}\,(t\,;\gamma) - D_t^\mu \underline{Q}\,(t\,;\gamma) \right) + D_t^\mu \underline{Q}\,(t\,;\gamma) = D_t^\mu \tilde{Q}\,(t\,;\gamma,\beta),$$

$$\beta \left(D_t^\mu \overline{L}\,(t\,;\gamma) - D_t^\mu \underline{L}\,(t\,;\gamma) \right) + D_t^\mu \underline{L}\,(t\,;\gamma) = D_t^\mu \tilde{L}\,(t\,;\gamma,\beta),$$

$$\beta \left(\overline{P}\,(t\,;\gamma) - \underline{P}\,(t\,;\gamma) \right) + \underline{P}\,(t\,;\gamma) = \tilde{P}\,(t\,;\gamma,\beta),$$

$$\beta \left(\overline{O}\,(t\,;\gamma) - \underline{O}\,(t\,;\gamma) \right) + \underline{O}\,(t\,;\gamma) = \tilde{O}\,(t\,;\gamma,\beta),$$

$$\beta \left(\overline{S}\,(t\,;\gamma) - \underline{S}\,(t\,;\gamma) \right) + \underline{S}\,(t\,;\gamma) = \tilde{S}\,(t\,;\gamma,\beta),$$

$$\beta \left(\overline{Q}\,(t\,;\gamma) - \underline{Q}\,(t\,;\gamma) \right) + \underline{Q}\,(t\,;\gamma) = \tilde{Q}\,(t\,;\gamma,\beta),$$

$$\beta \left(\overline{L}\,(t\,;\gamma) - \underline{L}\,(t\,;\gamma) \right) + \underline{L}\,(t\,;\gamma) = \tilde{L}\,(t\,;\gamma,\beta),$$

$$\beta \left(-0.02\gamma + 0.02 \right) + 0.01\gamma + \eta - 0.01 = \eta_0,$$

$$\beta \left(-0.002\gamma + 0.002 \right) + 0.001\gamma + \alpha_1 - 0.001 = \rho_1,$$

$$\beta \left(-0.0002\gamma + 0.0002 \right) + 0.0001\gamma + \alpha_2 - 0.0001 = \rho_2. \tag{5.9}$$

Substituting all the above equations in Eq. (5.8), we get

$$
\begin{cases}
D_t^\mu \tilde{P}\,(t\,;\gamma,\beta) = \lambda - b\,\tilde{P}\,(t\,;\gamma,\beta)\,\tilde{S}\,(t\,;\gamma,\beta) - \eta_0 \tilde{P}\,(t\,;\gamma,\beta), \\[4pt]
D_t^\mu \tilde{O}\,(t\,;\gamma,\beta) = b\tilde{P}\,(t\,;\gamma,\beta)\,\tilde{S}\,(t\,;\gamma,\beta) - \rho_1 \tilde{O}\,(t\,;\gamma,\beta) - \eta_0 \tilde{O}\,(t\,;\gamma,\beta), \\[4pt]
D_t^\mu \tilde{S}\,(t\,;\gamma,\beta) = \rho_1 \tilde{O}\,(t\,;\gamma,\beta) + \rho_2 \tilde{S}\,(t\,;\gamma,\beta)\,\tilde{Q}\,(t\,;\gamma,\beta) - (\eta_0 + r)\,\tilde{S}\,(t\,;\gamma,\beta), \\[4pt]
D_t^\mu \tilde{Q}\,(t\,;\gamma,\beta) = -\rho_2 \tilde{S}\,(t\,;\gamma,\beta)\,\tilde{Q}\,(t\,;\gamma,\beta) - \eta_0 \tilde{Q}\,(t\,;\gamma,\beta) + r\,(1-\delta)\,\tilde{S}\,(t\,;\gamma,\beta), \\[4pt]
D_t^\mu \tilde{L}\,(t\,;\gamma,\beta) = \delta r \tilde{S}\,(t\,;\gamma,\beta) - \eta_0 \tilde{L}\,(t\,;\gamma,\beta),
\end{cases} \tag{5.10}
$$

subject to initial conditions:

$$P\,(0) = 40, \quad O\,(0) = 10, \quad S\,(0) = 20, \quad Q\,(0) = 10, \quad \text{and} \quad L\,(0) = 5. \tag{5.11}$$

Using FRDTM, Eq. (5.10) reduces to the following expressions:

$$
\begin{cases}
\tilde{P}_{k+1}(\gamma,\beta) = \frac{\Gamma(1+\mu k)}{\Gamma(1+\mu k+\mu)} \\
\quad \left\{ \lambda\chi(k) - b\sum_{i=0}^{k} \tilde{P}_i(\gamma,\beta)\tilde{S}_{k-i}(\gamma,\beta) - \eta_0\tilde{P}_k(\gamma,\beta) \right\}, \\[2mm]
\tilde{O}_{k+1}(\gamma,\beta) = \frac{\Gamma(1+\mu k)}{\Gamma(1+\mu k+\mu)} \\
\quad \left\{ b\sum_{i=0}^{k} \tilde{P}_i(\gamma,\beta)\tilde{S}_{k-i}(\gamma,\beta) - \rho_1\tilde{O}_k(\gamma,\beta) - \eta_0\tilde{O}_k(\gamma,\beta) \right\}, \\[2mm]
\tilde{S}_{k+1}(\gamma,\beta) = \frac{\Gamma(1+\mu k)}{\Gamma(1+\mu k+\mu)} \\
\quad \left\{ \rho_1\tilde{O}_k(\gamma,\beta) + \rho_2\sum_{i=0}^{k} \tilde{S}_i(\gamma,\beta)\tilde{Q}_{k-i}(\gamma,\beta) - (\eta_0+r)\tilde{S}_k(\gamma,\beta) \right\}, \\[2mm]
\tilde{Q}_{k+1}(\gamma,\beta) = \frac{\Gamma(1+\mu k)}{\Gamma(1+\mu k+\mu)} \\
\quad \left\{ -\rho_2\sum_{i=0}^{k} \tilde{S}_i(\gamma,\beta)\tilde{Q}_{k-i}(\gamma,\beta) - \eta_0\tilde{Q}_k(\gamma,\beta) + r(1-\delta)\tilde{S}_k(\gamma,\beta) \right\}, \\[2mm]
\tilde{L}_{k+1}(\gamma,\beta) = \frac{\Gamma(1+\mu k)}{\Gamma(1+\mu k+\mu)} \\
\quad \left\{ \delta r\tilde{S}_k(\gamma,\beta) - \eta_0\tilde{L}_k(\gamma,\beta) \right\},
\end{cases}
\tag{5.12}
$$

where

$$
\chi(k) = \begin{cases} 1, & k=0 \\ 0, & k>0 \end{cases}
$$

Applying FRDTM to initial conditions (Eq. (5.11)), we have the following transformed initial conditions:

$$
\tilde{P}_0(\gamma,\beta) = 40, \quad \tilde{O}_0(\gamma,\beta) = 10, \quad \tilde{S}_0(\gamma,\beta) = 20, \quad \tilde{Q}_0(\gamma,\beta) = 10, \quad \text{and} \quad \tilde{L}_0(\gamma,\beta) = 5.
\tag{5.13}
$$

Substituting Eq. (5.13) into Eq. (5.12), we obtain the following expressions successively for $k = 0, 1, 2, \ldots$

$$
\tilde{P}_1(\gamma,\beta) = \frac{\lambda - 800b - 40\eta_0}{\Gamma(1+\mu)},
$$

$$
\tilde{O}_1(\gamma,\beta) = \frac{800b - 10\rho_1 - 10\eta_0}{\Gamma(1+\mu)},
$$

$$
\tilde{S}_1(\gamma,\beta) = \frac{10\rho_1 + 200\rho_2 - 20\eta_0 - 20r}{\Gamma(1+\mu)},
$$

$$
\tilde{Q}_1(\gamma,\beta) = \frac{-20\delta r + 20r - 10\eta_0 - 200\rho_2}{\Gamma(1+\mu)},
$$

$$\tilde{L}_1(\gamma, \beta) = \frac{20\delta r - 5\eta_0}{\Gamma(1 + \mu)},$$

$$\tilde{P}_2(\gamma, \beta) = \frac{\left(16000b^2 + (800r - 20\lambda + 2400\eta_0 - 400\rho_1 - 8000\rho_2)b - \eta_0(\lambda - 40\eta_0)\right)}{\Gamma(1 + 2\mu)},$$

$$\tilde{O}_2(\gamma, \beta) = \frac{\left(-16000b^2 + (-800r + 20\lambda - 2400\eta_0 - 400\rho_1 + 8000\rho_2)b + 10(\eta_0 + \rho_1)^2\right)}{\Gamma(1 + 2\mu)},$$

$$\tilde{S}_2(\gamma, \beta) = \frac{\left(-10\rho_1^2 + (800b - 10r - 20\eta_0 + 100\rho_2)\rho_1 - 2000\rho_2^2 \cdots\right.}{\Gamma(1 + 2\mu)}$$

$$\frac{\cdots + \rho_2(-400\delta r - 600\eta_0) + 20(\eta_0 + r)^2\right)}{\Gamma(1 + 2\mu)},$$

$$\tilde{Q}_2(\gamma, \beta) = \frac{\left((20\delta - 20)r^2 + \left(200\delta\rho_2 + 40\left(\eta_0 - \frac{\rho_1}{4}\right)(-1 + \delta)\right)r \cdots\right.}{\Gamma(1 + 2\mu)}$$

$$\frac{\cdots + 2000\rho_2^2 + (600\eta_0 - 100\rho_1)\rho_2 + 10\eta_0^2\right)}{\Gamma(1 + 2\mu)},$$

$$\tilde{L}_2(\gamma, \beta) = \frac{\left(-20r\left(r + 2\eta_0 - \frac{\rho_1}{2} - 10\rho_2\right)\delta + 5\eta_0^2\right)}{\Gamma(1 + 2\mu)}.$$

Continuing likewise, we may obtain all the values of $\{\tilde{P}_k, \tilde{O}_k, \tilde{S}_k, \tilde{Q}_k, \tilde{L}_k\}_{k=0}^{\infty}$. Taking inverse differential transform to $\{\tilde{P}_k, \tilde{O}_k, \tilde{S}_k, \tilde{Q}_k, \tilde{L}_k\}_{k=0}^{\infty}$, the following nth order approximate solutions are obtained:

$$\begin{cases} \tilde{P}_n(t, \gamma, \beta) = \sum_{k=0}^{n} \tilde{P}_n(\gamma, \beta) t^{\mu k}, \\ \tilde{O}_n(t, \gamma, \beta) = \sum_{k=0}^{n} \tilde{O}_n(\gamma, \beta) t^{\mu k}, \\ \tilde{S}_n(t, \gamma, \beta) = \sum_{k=0}^{n} \tilde{S}_n(\gamma, \beta) t^{\mu k}, \\ \tilde{Q}_n(t, \gamma, \beta) = \sum_{k=0}^{n} \tilde{Q}_n(\gamma, \beta) t^{\mu k}, \\ \tilde{L}_n(t, \gamma, \beta) = \sum_{k=0}^{n} \tilde{L}_n(\gamma, \beta) t^{\mu k}. \end{cases} \tag{5.14}$$

The lower- and upper-bound solutions can be evaluated by substituting $\beta = 0$ and $\beta = 1$ into Eq. (5.14), respectively. The above series solutions converge rapidly, and the rapid convergence solutions mean only a few terms are required for obtaining closed-form solutions. In particular, the solutions of this model ($n = 3$) for the integer-order derivative ($\mu = 1$) are as follows:

$$
\begin{cases}
\tilde{P}\,(t; \gamma = 1) = 40 - 113.0t + 207.169t^2 - 297.2381127t^3, \\
\tilde{O}\,(t; \gamma = 1) = 10 + 111.48t - 207.24248t^2 + 297.377499t^3, \\
\tilde{S}\,(t; \gamma = 1) = 20 - 16.48t + 7.24448t^2 - 2.421033653t^3, \\
\tilde{Q}\,(t; \gamma = 1) = 10 + 13.4t - 6.3968t^2 + 2.0755578671t^3, \\
\tilde{L}\,(t; \gamma = 1) = 5 + 1.35t - 0.69295t^2 + 0.2047353t^3.
\end{cases}
\tag{5.15}
$$

Similarly, one may consider other parameters involved in the model as well as initial conditions as uncertain, and accordingly, the solution of the model can be evaluated.

5.5 NUMERICAL RESULTS AND DISCUSSION

In this section, we have presented the solution of the uncertain time-fractional SEM using FRDTM. Here, all the numerical computations are done by truncating the infinite series solution to a finite number terms of solution ($n = 3$). It is a gigantic task to give all the results with respect to various parameters and constants involved in the titled problem. As such, a few essential results are reported in this study. In this numerical simulation, we have considered the values of the parameters [9, 10] as $\lambda = 1$, $b = 014$, $\eta = 0.05$, $\delta = 0.1$, $r = 0.8$, $\alpha_1 = 0.002$, and $\alpha_2 = 0.0025$. Numerical results of the titled model are depicted in Table 5.1 by varying t from 0 to 1, and γ from 0 to 1 keeping the fractional parameter μ as constant. Also, crisp results obtained by Abdullah et al. [10] are reflected in Table 5.1. It may be concluded that the crisp results lie in between lower-bound and upper-bound solutions of the model. It is interesting to note that both the lower-bound and upper-bound fuzzy solutions are equal to each other at $\gamma = 1$, which is the same as the solution of Abdullah et al. [10]. Fuzzy and interval solutions of the time-fractional SEM are illustrated graphically in Figs. 5.1 and 5.2. Figure 5.3 reflects the solution plots of the SEM at various values of fractional-order derivatives ($\mu = 0.3,\ 0.5,\ 0.7,\ \text{and}\ 0.9$).

Table 5.1: Fuzzy solutions of the model for integer-order derivative ($\mu = 1$)

$t \rightarrow$		0	0.4	0.8	1.0
$\gamma = 0$	$[\underline{P}, \overline{P}]$	[40, 40]	[9.0699, 8.7768]	[-68.758, -71.246]	[-160.47, -165.68]
	$[\underline{O}, \overline{O}]$	[10, 10]	[40.524, 40.407]	[117.96, 119.66]	[209.495, 213.760]
	$[\underline{S}, \overline{S}]$	[20, 20]	[14.457, 14.367]	[10.297, 10.127]	[8.46602, 8.21923]
	$[\underline{Q}, \overline{Q}]$	[10, 10]	[14.532, 14.406]	[17.822, 17.556]	[19.2416, 18.9185]
	$[\underline{L}, \overline{L}]$	[5, 5]	[5.4637, 5.4208]	[5.7852, 5.6978]	[5.91626, 5.80791]
$\gamma = 0.3$	$[\underline{P}, \overline{P}]$	[40, 40]	[9.0261, 8.8209]	[-69.129, -70.870]	[-161.24, -164.89]
	$[\underline{O}, \overline{O}]$	[10, 10]	[40.506, 40.424]	[118.214, 119.403]	[210.128, 213.113]
	$[\underline{S}, \overline{S}]$	[20, 20]	[14.443, 14.380]	[10.272, 10.1530]	[8.42942, 8.2566]
	$[\underline{Q}, \overline{Q}]$	[10, 10]	[14.513, 14.425]	[17.7824, 17.595]	[19.192, 18.966]
	$[\underline{L}, \overline{L}]$	[5, 5]	[5.4572, 5.4272]	[5.7720, 5.7108]	[5.8998, 5.8240]
$\gamma = 0.6$	$[\underline{P}, \overline{P}]$	[40, 40]	[8.9823, 8.8651]	[-69.5010, -70.495]	[-162.027, -164.113]
	$[\underline{O}, \overline{O}]$	[10, 10]	[40.488, 40.442]	[118.4673, 119.146]	[210.7640, 212.470]
	$[\underline{S}, \overline{S}]$	[20, 20]	[14.430, 14.394]	[10.2469, 10.1787]	[8.3926, 8.2939]
	$[\underline{Q}, \overline{Q}]$	[10, 10]	[14.494, 14.444]	[17.7421, 17.6355]	[19.14359, 19.0143]
	$[\underline{L}, \overline{L}]$	[5, 5]	[5.4508, 5.4336]	[5.7588, 5.72388]	[5.88350, 5.84016]
$\gamma = 1.0$	$[\underline{P}, \overline{P}]$	[40, 40]	[8.9238, 8.9238]	[-69.9977, -69.9977]	[-163.069, -163.069]
	$[\underline{O}, \overline{O}]$	[10, 10]	[40.465, 40.465]	[118.806, 118.806]	[211.605, 211.615]
	$[\underline{S}, \overline{S}]$	[20, 20]	[14.412, 14.412]	[10.2128, 10.2128]	[8.34344, 8.34344]
	$[\underline{Q}, \overline{Q}]$	[10, 10]	[14.469, 14.469]	[17.6887, 17.68787]	[19.0787, 19.0787]
	$[\underline{L}, \overline{L}]$	[5, 5]	[5.4422, 5.4422]	[5.74133, 5.74133]	[5.86178, 5.86178]
Reference [10]					
	$P(t)$	40	8.9238	-69.9977	-163.069
	$O(t)$	10	40.465	118.806	211.605
	$S(t)$	20	14.412	10.2128	8.34344
	$Q(t)$	10	14.469	17.6887	19.0787
	$L(t)$	5	5.4422	5.74133	5.86178

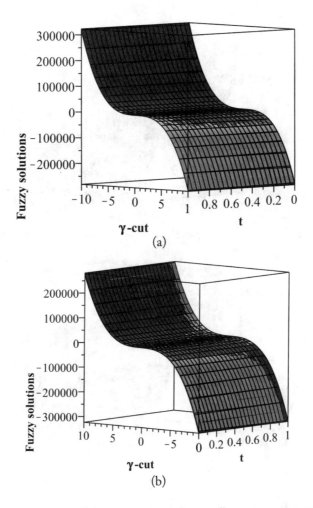

Figure 5.1: Fuzzy solution of time-fractional SEM for (a) $\tilde{P}\,(t;\gamma)$ and (b) $\tilde{O}\,(t;\gamma)$, at $\gamma \in [0, 1]$ and $\mu = 1$. (*Continues.*)

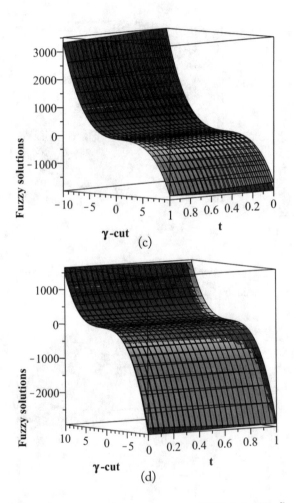

Figure 5.1: (*Continued.*) Fuzzy solution of time-fractional SEM for (c) $\tilde{S}\,(t;\gamma)$ and (d) $\tilde{Q}\,(t;\gamma)$, at $\gamma \in [0,1]$ and $\mu = 1$. (*Continues.*)

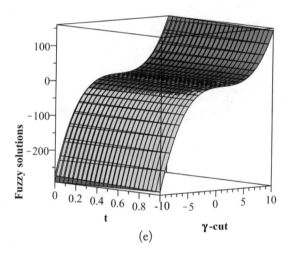

(e)

Figure 5.1: (*Continued.*) Fuzzy solution of time-fractional SEM for (e) $\tilde{L}\left(t;\gamma\right)$, at $\gamma \in [0,1]$ and $\mu = 1$.

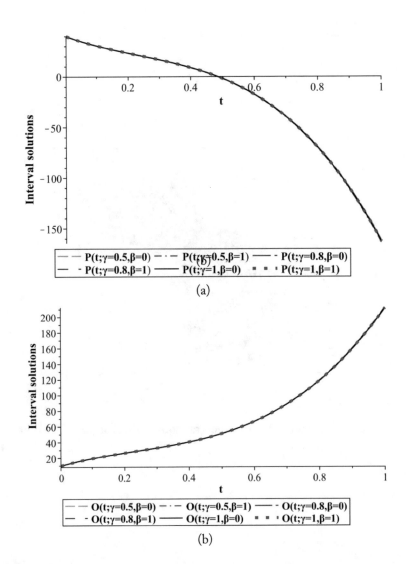

(a)

(b)

Figure 5.2: Interval solution of titled model for (a) $\tilde{P}(t; \gamma)$ and (b) $\tilde{O}(t; \gamma)$ at $\gamma \in [0, 1]$ and $\mu = 1$. (*Continues.*)

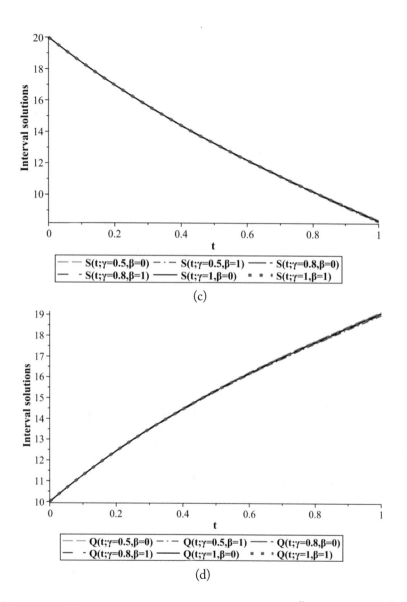

(c)

(d)

Figure 5.2: (*Continued.*) Interval solution of titled model for (c) $\tilde{S}(t;\gamma)$ and (d) $\tilde{Q}(t;\gamma)$ at $\gamma \in$ [0, 1] and $\mu = 1$. (*Continues.*)

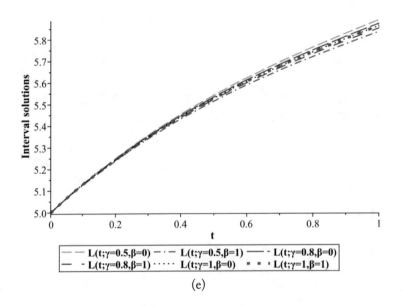

(e)

Figure 5.2: (*Continued.*) Interval solution of titled model for (e) $\tilde{L}(t;\gamma)$ at $\gamma \in [0,1]$ and $\mu = 1$.

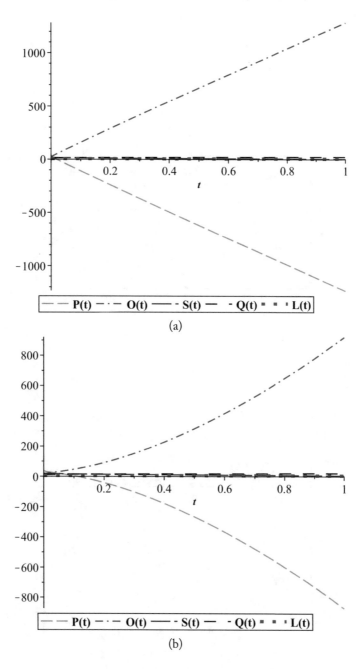

Figure 5.3: Solution plots of fractional SEM for different fractional order derivatives (a) $\mu = 0.3$ and (b) $\mu = 0.5$ at $\gamma = 1$ (crisp result). (*Continues.*)

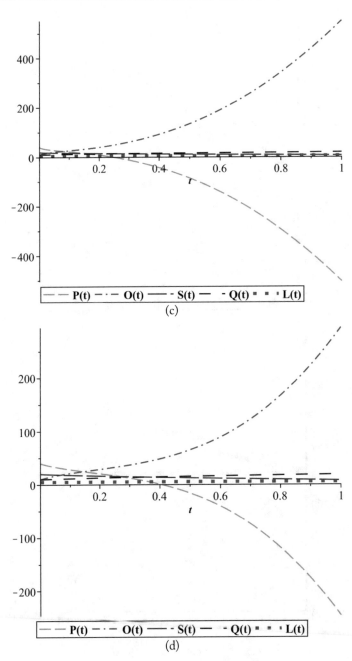

Figure 5.3: (*Continued.*) Solution plots of fractional SEM for different fractional order derivatives (c) $\mu = 0.7$ and (d) $\mu = 0.9$ at $\gamma = 1$ (crisp result).

5.6   REFERENCES

[1] D. Bernoulli. Essaid'une nouvelle analyse de la mortalite cause par la petite verole. *Mémoires de Mathématique et de Physique Académie Royale des Sciences*, pp. 1–41, 1766. 55

[2] R. Ross, H. E. Annett, and E. E. Austen. Report of the Malaria expedition of the Liverpool school of tropical medicine and medical parasitology, pp. 1–71, University Press of Liverpool, Liverpool, 1900. DOI: 10.5962/bhl.title.99657. 55

[3] F. J. Santonja, E. Sanchez, M. Rubio, and J. L. Morera. Alcohol consumption in Spain and its economic costs: A mathematical modeling approach. *Mathematical and Computer Modelling*, 52:999–1003, 2010. DOI: 10.1016/j.mcm.2010.02.029. 55

[4] F. J. Santonja, R. J. Villanueva, L. Jodar, and G. Gonzalez. Mathematical modeling of social obesity epidemic in the region of Valencia, Spain. *Mathematical and Computer Modelling of Dynamical Systems*, 16(1):23–34, 2010. DOI: 10.1080/13873951003590149. 55

[5] E. Sanchez, R. J. Villanueva, F. J. Santonja, and M. Rubio. Predicting cocaine consumption in Spain. A mathematical modeling approach. *Drugs: Education Prevention and Policy*, 18(2):105–108, 2011. DOI: 10.3109/09687630903443299. 55

[6] F. Guerrero, F. J. Santonja, and R. J. Villanueva. Analysing the effect of Spanish smoke-free legislation of year 2006: A new method to quantify its impact using a dynamic model. *International Journal of Drug Policy*, 22:247–251, 2011. DOI: 10.1016/j.drugpo.2011.05.003. 55

[7] WHO, 2010. http://www.emro.who.int/tfi/facts.htm#fact2 55

[8] C. Castillo-Garsow, G. Jordan-Salivia, and A. R. Herrera. Mathematical models for the dynamics of tobacco use, recovery, and relapse. *Technical Report Series BU-1505-M*, Cornell University, Ithaca, NY, 1997.

[9] P. Veeresha, D. G. Prakasha, and H. M. Baskonus. Solving smoking epidemic model of fractional order using a modified homotopy analysis transform method. *Mathematical Sciences*, 2019. https://doi.org/10.1007/s40096--019-0284-6 DOI: 10.1007/s40096-019-0284-6. 56, 63

[10] M. Abdullah, A. Ahmad, N. Raza, M. Farman, and M. O. Ahmad. Approximate solution and analysis of smoking epidemic model with caputo fractional derivatives, *International Journal of Applied Computational Mathematics*, 4:112–128, 2018. DOI: 10.1007/s40819-018-0543-5. 56, 63

[11] R. Ullah, M. Khan, and G. Zaman. Dynamical features of a mathematical model on smoking. *Journal of Applied Environmental and Biological Sciences*, 6(1):92–96, 2016. 56

[12] R. M. Jena, S. Chakraverty, and D. Baleanu. On the solution of imprecisely defined nonlinear time-fractional dynamical model of marriage. *Mathematics*, 7:689–704, 2019. DOI: 10.3390/math7080689.

[13] R. M. Jena, S. Chakraverty, and D. Baleanu. On new solutions of time-fractional wave equations arising in Shallow water wave propagation. *Mathematics*, 7:722–734, 2019. DOI: 10.3390/math7080722.

[14] R. M. Jena and S. Chakraverty. Solving time-fractional Navier–Stokes equations using homotopy perturbation Elzaki transform. *SN Applied Sciences*, 1(1):16, 2019. DOI: 10.1007/s42452-018-0016-9.

[15] R. M. Jena and S. Chakraverty. Residual power series method for solving time-fractional model of vibration equation of large membranes. *Journal of Applied and Computational Mechanics*, 5:603–615, 2019. DOI: 10.22055/jacm.2018.26668.1347.

[16] R. M. Jena and S. Chakraverty. A new iterative method based solution for fractional Black–Scholes option pricing equations (BSOPE). *SN Applied Sciences*, 1:95–105, 2019. DOI: 10.1007/s42452-018-0106-8.

[17] R. M. Jena and S. Chakraverty. Analytical solution of Bagley–Torvik equations using Sumudu transformation method. *SN Applied Sciences*, 1(3):246, 2019. DOI: 10.1007/s42452-019-0259-0.

[18] R. M. Jena, S. Chakraverty, and S. K. Jena. Dynamic response analysis of fractionally damped beams subjected to external loads using homotopy analysis method. *Journal of Applied and Computational Mechanics*, 5:355–366, 2019. DOI: 10.22055/jacm.2019.27592.1419.

[19] S. Chakraverty, S. Tapaswini, and D. Behera, *Fuzzy Arbitrary Order System: Fuzzy Fractional Differential Equations and Applications*. John Wiley & Sons, 2016. DOI: 10.1002/9781119004233.

[20] S. Chakraverty, S. Tapaswini, and D. Behera. *Fuzzy Differential Equations and Applications for Engineers and Scientists*. Taylor & Francis Group, CRC Press, Boca Raton, FL, 2016. DOI: 10.1201/9781315372853.

[21] S. Chakraverty, D. M. Sahoo, and N. R. Mahato. *Concepts of Soft Computing: Fuzzy and ANN with Programming*. Springer, Singapore, 2019. DOI: 10.1007/978-981-13-7430-2.

CHAPTER 6

Time-Fractional Model of HIV-I Infection of CD4$^+$ T Lymphocyte Cells in Uncertain Environment

6.1 INTRODUCTION

Human immunodeficiency virus (HIV) is a retrovirus (a type of RNA virus that inserts a copy of its genome into the DNA of a host cell that it invades, thus changing the genome of that cell) that causes acquired immune deficiency syndrome (AIDS) [1]. HIV infects, damages, and reduces CD4$^+$ T-cells. Therefore, it decreases the resistance of the immune system [2]. Living beings become more sensitive to infections and lose their protection. AIDS is one of the most dangerous diseases of our time. The CD4$^+$ T lymphocyte cells are the white blood cells that play a crucial role in protecting the human body from infection by improving the immune system [3].

The main target of this chapter is to investigate the solution of a time-fractional model of HIV-I infection of CD4$^+$ T lymphocyte cells using FRDTM in an uncertain environment. In this model, initial conditions are considered as the fuzzy triangular numbers.

6.2 MATHEMATICAL MODEL WITH FUZZY INITIAL CONDITIONS

First, the fractional fuzzy HIV-I model is converted to an interval-based fuzzy differential equation by using the single parametric form [18, 19, 25–27]. Then, by using the double parametric form [25–27], the interval-based fuzzy differential equation is transformed into a fractional HIV-I model having two parameters which may control the uncertainty. Finally, FRDTM is used to solve the corresponding fractional model for obtaining the required solution in terms of interval/fuzzy.

Let us consider the fuzzy time-fractional model of HIV-I infection of CD4$^+$ T lymphocyte cells, which is as follows:

$$\begin{cases} \frac{d^{\alpha_1}\tilde{\psi}(t)}{dt^{\alpha_1}} = s' - \mu\tilde{\psi}(t) - b\tilde{\psi}(t)\tilde{\zeta}(t), \\[2mm] \frac{d^{\alpha_2}\tilde{\xi}(t)}{dt^{\alpha_2}} = b\tilde{\psi}(t)\tilde{\zeta}(t) - \varepsilon\tilde{\xi}(t), \qquad\qquad \text{for } 0 < \alpha_i \leq 1, \ i = 1, 2, 3 \\[2mm] \frac{d^{\alpha_3}\tilde{\zeta}(t)}{dt^{\alpha_3}} = c\tilde{\xi}(t) - r\tilde{\zeta}(t), \end{cases} \tag{6.1}$$

with the fuzzy initial conditions:

$$\tilde{\psi}(0) = (a_1, 100, c_1), \quad \tilde{\xi}(0) = (a_2, 0, c_2), \quad \text{and } \tilde{\zeta}(0) = (a_3, 1, c_3). \tag{6.2}$$

It may be noted that the crisp form of the model is given in references [6, 8–10]. In the following sections, the solutions of the title model have been discussed considering the initial conditions as a TFN in three different cases.

Case I. Let us consider the Eq. (6.1) with initial conditions as follows:

$$\tilde{\psi}(0) = (99.5, 100, 100.5), \quad \xi(0) = 0, \quad \text{and } \zeta(0) = 1. \tag{6.3}$$

Now, applying single parametric form, Eqs. (6.1) and (6.3), reduce to the following form:

$$\begin{cases} \left[\frac{d^{\alpha_1}\underline{\psi}(t;\gamma)}{dt^{\alpha_1}}, \frac{d^{\alpha_1}\overline{\psi}(t;\gamma)}{dt^{\alpha_1}}\right] = s' - \mu\left[\underline{\psi}(t;\gamma), \overline{\psi}(t;\gamma)\right] \\[2mm] \qquad\qquad -b\left[\underline{\psi}(t;\gamma), \overline{\psi}(t;\gamma)\right]\left[\underline{\zeta}(t;\gamma), \overline{\zeta}(t;\gamma)\right], \\[2mm] \left[\frac{d^{\alpha_2}\underline{\xi}(t;\gamma)}{dt^{\alpha_2}}, \frac{d^{\alpha_2}\overline{\xi}(t;\gamma)}{dt^{\alpha_2}}\right] = b\left[\underline{\psi}(t;\gamma), \overline{\psi}(t;\gamma)\right]\left[\underline{\zeta}(t;\gamma), \overline{\zeta}(t;\gamma)\right] \\[2mm] \qquad\qquad -\varepsilon\left[\underline{\xi}(t;\gamma), \overline{\xi}(t;\gamma)\right], \\[2mm] \left[\frac{d^{\alpha_3}\underline{\zeta}(t;\gamma)}{dt^{\alpha_3}}, \frac{d^{\alpha_3}\overline{\zeta}(t;\gamma)}{dt^{\alpha_3}}\right] = c\left[\underline{\xi}(t;\gamma), \overline{\xi}(t;\gamma)\right] \\[2mm] \qquad\qquad -r\left[\underline{\zeta}(t;\gamma), \overline{\zeta}(t;\gamma)\right], \end{cases} \tag{6.4}$$

subject to fuzzy initial conditions (in term of single parametric form):

$$\tilde{\psi}(0;\gamma) = (0.5\gamma + 99.5, -0.5\gamma + 100.5), \ \xi(0) = 0, \ \text{and } \zeta(0) = 1, \ \text{where } \gamma \in [0,1]. \tag{6.5}$$

By applying the double parametric form to Eqs. (6.4) and (6.5), the following expressions are obtained:

$$\begin{cases} \beta \left(\dfrac{d^{\alpha_1} \overline{\psi}(t\,;\gamma)}{dt^{\alpha_1}} - \dfrac{d^{\alpha_1} \underline{\psi}(t\,;\gamma)}{dt^{\alpha_1}} \right) + \dfrac{d^{\alpha_1} \underline{\psi}(t\,;\gamma)}{dt^{\alpha_1}} = s' - \mu \left\{ \beta \left(\overline{\psi}(t\,;\gamma) - \underline{\psi}(t\,;\gamma) \right) + \underline{\psi}(t\,;\gamma) \right\} \\ \quad -b \left\{ \beta \left(\overline{\psi}(t\,;\gamma) - \underline{\psi}(t\,;\gamma) \right) + \underline{\psi}(t\,;\gamma) \right\} \left\{ \beta \left(\overline{\xi}(t\,;\gamma) - \underline{\xi}(t\,;\gamma) \right) + \underline{\xi}(t\,;\gamma) \right\}, \\ \beta \left(\dfrac{d^{\alpha_2} \overline{\xi}(t\,;\gamma)}{dt^{\alpha_2}} - \dfrac{d^{\alpha_2} \underline{\xi}(t\,;\gamma)}{dt^{\alpha_2}} \right) + \dfrac{d^{\alpha_2} \underline{\xi}(t\,;\gamma)}{dt^{\alpha_2}} = b \left\{ \beta \left(\overline{\psi}(t\,;\gamma) - \underline{\psi}(t\,;\gamma) \right) + \underline{\psi}(t\,;\gamma) \right\} \\ \quad \left\{ \beta \left(\overline{\xi}(t\,;\gamma) - \underline{\xi}(t\,;\gamma) \right) + \underline{\xi}(t\,;\gamma) \right\} - \varepsilon \left\{ \beta \left(\overline{\xi}(t\,;\gamma) - \underline{\xi}(t\,;\gamma) \right) + \underline{\xi}(t\,;\gamma) \right\}, \\ \beta \left(\dfrac{d^{\alpha_3} \overline{\zeta}(t\,;\gamma)}{dt^{\alpha_3}} - \dfrac{d^{\alpha_3} \underline{\zeta}(t\,;\gamma)}{dt^{\alpha_3}} \right) + \dfrac{d^{\alpha_3} \underline{\zeta}(t\,;\gamma)}{dt^{\alpha_3}} = c \left\{ \beta \left(\overline{\xi}(t\,;\gamma) - \underline{\xi}(t\,;\gamma) \right) + \underline{\xi}(t\,;\gamma) \right\} \\ \quad -r \left\{ \beta \left(\overline{\xi}(t\,;\gamma) - \underline{\xi}(t\,;\gamma) \right) + \underline{\xi}(t\,;\gamma) \right\}, \end{cases}$$

(6.6)

subject to fuzzy initial conditions (in terms of DPF):

$$\tilde{\psi}(0\,;\gamma, \beta) = [\beta(-\gamma + 1) + 0.5\gamma + 99.5], \ \xi(0) = 0, \text{ and } \zeta(0) = 1, \text{ where } \beta, \gamma \in [0, 1]. \quad (6.7)$$

Let us denote

$$\beta \left(\frac{d^{\alpha_1} \overline{\psi}(t\,;\gamma)}{dt^{\alpha_1}} - \frac{d^{\alpha_1} \underline{\psi}(t\,;\gamma)}{dt^{\alpha_1}} \right) + \frac{d^{\alpha_1} \underline{\psi}(t\,;\gamma)}{dt^{\alpha_1}} = \frac{d^{\alpha_1} \tilde{\psi}(t\,;\gamma, \beta)}{dt^{\alpha_1}},$$

$$\beta \left(\frac{d^{\alpha_2} \overline{\xi}(t\,;\gamma)}{dt^{\alpha_2}} - \frac{d^{\alpha_2} \underline{\xi}(t\,;\gamma)}{dt^{\alpha_2}} \right) + \frac{d^{\alpha_2} \underline{\xi}(t\,;\gamma)}{dt^{\alpha_2}} = \frac{d^{\alpha_2} \tilde{\xi}(t\,;\gamma, \beta)}{dt^{\alpha_2}},$$

$$\beta \left(\frac{d^{\alpha_3} \overline{\zeta}(t\,;\gamma)}{dt^{\alpha_3}} - \frac{d^{\alpha_3} \underline{\zeta}(t\,;\gamma)}{dt^{\alpha_3}} \right) + \frac{d^{\alpha_3} \underline{\zeta}(t\,;\gamma)}{dt^{\alpha_3}} = \frac{d^{\alpha_3} \tilde{\zeta}(t\,;\gamma, \beta)}{dt^{\alpha_3}},$$

$$\beta \left(\overline{\psi}(t\,;\gamma) - \underline{\psi}(t\,;\gamma) \right) + \underline{\psi}(t\,;\gamma) = \tilde{\psi}(t\,;\gamma, \beta),$$

$$\beta \left(\overline{\xi}(t\,;\gamma) - \underline{\xi}(t\,;\gamma) \right) + \underline{\xi}(t\,;\gamma) = \tilde{\xi}(t\,;\gamma, \beta),$$

$$\beta \left(\overline{\zeta}(t\,;\gamma) - \underline{\zeta}(t\,;\gamma) \right) + \underline{\zeta}(t\,;\gamma) = \tilde{\zeta}(t\,;\gamma, \beta),$$

$$\beta(-\gamma + 1) + 0.5\gamma + 99.5 = \eta.$$

Substituting all the above equations into Eqs. (6.6) and (6.7), we have

$$\begin{cases} \dfrac{d^{\alpha_1} \tilde{\psi}(t\,;\gamma, \beta)}{dt^{\alpha_1}} = s' - \mu \tilde{\psi}(t\,;\gamma, \beta) - b\tilde{\psi}(t\,;\gamma, \beta) \ \tilde{\xi}(t\,;\gamma, \beta), \\ \dfrac{d^{\alpha_2} \tilde{\xi}(t\,;\gamma, \beta)}{dt^{\alpha_2}} = b\,\tilde{\psi}(t\,;\gamma, \beta) \ \tilde{\xi}(t\,;\gamma, \beta) - \varepsilon \tilde{\xi}(t\,;\gamma, \beta), \\ \dfrac{d^{\alpha_3} \tilde{\zeta}(t\,;\gamma, \beta)}{dt^{\alpha_3}} = c\,\tilde{\xi}(t\,;\gamma, \beta) - r\,\tilde{\zeta}(t\,;\gamma, \beta). \end{cases} \quad (6.8)$$

$$\tilde{\psi}(0;\gamma,\beta) = \eta, \ \xi(0) = 0, \ \text{and} \ \zeta(0) = 1. \tag{6.9}$$

Now, applying FRDTM to Eqs. (6.8) and (6.9), we obtain the following expression:

$$
\begin{cases}
\tilde{\psi}_{k+1}(\gamma,\beta) = \frac{\Gamma(1+\alpha_1 k)}{\Gamma(1+\alpha_1 k+\alpha_1)}\left\{s'\delta(k) - \mu\tilde{\psi}_k(\gamma,\beta) - b\sum_{i=0}^{k}\tilde{\psi}_i(\gamma,\beta)\tilde{\zeta}_{k-i}(\gamma,\beta)\right\}, \\
\tilde{\xi}_{k+1}(\gamma,\beta) = \frac{\Gamma(1+\alpha_2 k)}{\Gamma(1+\alpha_2 k+\alpha_2)}\left\{b\sum_{i=0}^{k}\tilde{\psi}_i(\gamma,\beta)\tilde{\zeta}_{k-i}(\gamma,\beta) - \varepsilon\tilde{\xi}_k(\gamma,\beta)\right\}, \\
\tilde{\zeta}_{k+1}(\gamma,\beta) = \frac{\Gamma(1+\alpha_3 k)}{\Gamma(1+\alpha_3 k+\alpha_3)}\left\{c\tilde{\xi}_k(\gamma,\beta) - r\tilde{\zeta}(\gamma,\beta)\right\},
\end{cases}
$$

$$\tag{6.10}$$

subject to transformed initial conditions:

$$\tilde{\psi}_0(\gamma,\beta) = \eta, \ \tilde{\xi}_0(0) = 0 \ \text{and} \ \tilde{\zeta}_0(0) = 1 \ \text{for} \ k = 0,1,2,\dots \tag{6.11}$$

where

$$\delta(k) = \begin{cases} 1, & k = 0, \\ 0, & k \neq 0. \end{cases}$$

Plugging initial conditions Eq. (6.11) into Eq. (6.10) for $k = 0,1,2,\dots$ the following equations are obtained successively:

$$\tilde{\psi}_1(\gamma,\beta) = \frac{1}{\Gamma(1+\alpha_1)}(s' - \mu\eta - b\eta),$$

$$\tilde{\xi}_1(\gamma,\beta) = \frac{b\eta}{\Gamma(1+\alpha_2)},$$

$$\tilde{\zeta}_1(\gamma,\beta) = \frac{-r}{\Gamma(1+\alpha_3)},$$

$$\tilde{\psi}_2(\gamma,\beta) = \frac{\Gamma(1+\alpha_1)}{\Gamma(1+2\alpha_1)}\left[\left(-\frac{\mu(-b\eta - \mu\eta + s')}{\Gamma(1+\alpha_1)} - b\left(\frac{-r\eta}{\Gamma(1+\alpha_3)} + \frac{(-b\eta - \mu\eta + s')}{\Gamma(1+\alpha_1)}\right)\right)\right],$$

$$\tilde{\xi}_2(\gamma,\beta) = \frac{\Gamma(1+\alpha_2)}{\Gamma(1+2\alpha_2)}\left[b\left(\frac{-r\eta}{\Gamma(1+\alpha_3)} + \frac{(-b\eta - \mu\eta + s')}{\Gamma(1+\alpha_1)}\right) - \frac{\varepsilon b\eta}{\Gamma(1+\alpha_2)}\right],$$

$$\tilde{\zeta}_2(\gamma,\beta) = \frac{\Gamma(1+\alpha_3)}{\Gamma(1+2\alpha_3)}\left[\frac{r^2}{\Gamma(1+\alpha_3)} + \frac{cb\eta}{\Gamma(1+\alpha_2)}\right],$$

$$\tilde{\psi}_3(\gamma,\beta) = \frac{\Gamma(1+2\alpha_1)}{\Gamma(1+3\alpha_1)}$$

$$
\begin{pmatrix}
-\mu\left(\frac{\Gamma(1+\alpha_1)}{\Gamma(1+2\alpha_1)}\left(\left(-\frac{\mu(-b\eta-\mu\eta+s')}{\Gamma(1+\alpha_1)} - b\left(\frac{-r\eta}{\Gamma(1+\alpha_3)} + \frac{(-b\eta-\mu\eta+s')}{\Gamma(1+\alpha_1)}\right)\right)\right)\right) - \\
b\begin{pmatrix}\eta\left(\frac{\Gamma(1+\alpha_3)}{\Gamma(1+2\alpha_3)}\left[\frac{r^2}{\Gamma(1+\alpha_3)} + \frac{cb\eta}{\Gamma(1+\alpha_2)}\right]\right) - \left(\frac{-r}{\Gamma(1+\alpha_1)\Gamma(1+\alpha_3)}(s'-\mu\eta-b\eta)\right) \\ -\left(\frac{\Gamma(1+\alpha_1)}{\Gamma(1+2\alpha_1)}\left[-\frac{\mu(-b\eta-\mu\eta+s')}{\Gamma(1+\alpha_1)} - b\left(\frac{-r\eta}{\Gamma(1+\alpha_3)} + \frac{(-b\eta-\mu\eta+s')}{\Gamma(1+\alpha_1)}\right)\right]\right)\end{pmatrix}
\end{pmatrix}
$$

$$\tilde{\xi}_3(\gamma, \beta) = \frac{\Gamma(1+2\alpha_2)}{\Gamma(1+3\alpha_2)}$$

$$\left(\begin{array}{l} -\varepsilon \frac{\Gamma(1+\alpha_2)}{\Gamma(1+2\alpha_2)} \left[b \left(\frac{-r\eta}{\Gamma(1+\alpha_3)} + \frac{(-b\eta-\mu\eta+s')}{\Gamma(1+\alpha_1)} \right) - \frac{\varepsilon b\eta}{\Gamma(1+\alpha_2)} \right] + \\ b \left(\begin{array}{l} \eta \left(\frac{\Gamma(1+\alpha_3)}{\Gamma(1+2\alpha_3)} \left[\frac{r^2}{\Gamma(1+\alpha_3)} + \frac{cb\eta}{\Gamma(1+\alpha_2)} \right] \right) - \left(\frac{-r}{\Gamma(1+\alpha_1)\Gamma(1+\alpha_3)} (s' - \mu\eta - b\eta) \right) \\ - \left(\frac{\Gamma(1+\alpha_1)}{\Gamma(1+2\alpha_1)} \left[\left(-\frac{\mu(-b\eta-\mu\eta+s')}{\Gamma(1+\alpha_1)} - b \left(\frac{-r\eta}{\Gamma(1+\alpha_3)} + \frac{(-b\eta-\mu\eta+s')}{\Gamma(1+\alpha_1)} \right) \right) \right] \right) \end{array} \right) \end{array} \right)$$

$$\tilde{\zeta}_3(\gamma, \beta) = \frac{\Gamma(1+2\alpha_3)}{\Gamma(1+3\alpha_3)}$$

$$\left(\begin{array}{l} c \frac{\Gamma(1+\alpha_2)}{\Gamma(1+2\alpha_2)} \left[b \left(\frac{-r\eta}{\Gamma(1+\alpha_3)} + \frac{(-b\eta-\mu\eta+s')}{\Gamma(1+\alpha_1)} \right) - \frac{\varepsilon(b\eta)}{\Gamma(1+\alpha_2)} \right] - r \frac{\Gamma(1+\alpha_3)}{\Gamma(1+2\alpha_3)} \\ \left[\frac{r^2}{\Gamma(1+\alpha_3)} + \frac{b\eta c}{\Gamma(1+\alpha_2)} \right] \end{array} \right).$$

Continuing the procedure likewise, we may get all the values of $\{\psi_k\}_{k=0}^{\infty}$, $\{\xi_k\}_{k=0}^{\infty}$, and $\{\zeta_k\}_{k=0}^{\infty}$. Now, applying the inverse differential transform to $\{\psi_k\}_{k=0}^{\infty}$, $\{\xi_k\}_{k=0}^{\infty}$, and $\{\zeta_k\}_{k=0}^{\infty}$, we obtain the following nth order approximate solutions:

$$\begin{cases} \tilde{\psi}_n(t; \gamma, \beta) = \sum_{k=0}^{n} \tilde{\psi}_k(\gamma, \beta) \, t^{\alpha_1 k}, \\ \tilde{\xi}_n(t; \gamma, \beta) = \sum_{k=0}^{n} \tilde{\xi}_k(\gamma, \beta) \, t^{\alpha_2 k}, \\ \tilde{\zeta}_n(t; \gamma, \beta) = \sum_{k=0}^{n} \tilde{\zeta}_k(\gamma, \beta) \, t^{\alpha_3 k}. \end{cases} \tag{6.12}$$

The exact solution of this model may be written as

$$\begin{cases} \tilde{\psi}(t; \gamma, \beta) = \lim_{n \to \infty} \tilde{\psi}_n(t; \gamma, \beta), \\ \tilde{\xi}(t; \gamma, \beta) = \lim_{n \to \infty} \tilde{\xi}_n(t; \gamma, \beta), \\ \tilde{\zeta}(t; \gamma, \beta) = \lim_{n \to \infty} \tilde{\zeta}_n(t; \gamma, \beta). \end{cases} \tag{6.13}$$

To obtain the lower- and upper-bound solutions of the model, we need to substitute $\beta = 0$ and $\beta = 1$, respectively. Mathematically, one may respectively write the following equations for the same:

$$\tilde{\psi}(t; \gamma, 0,) = \underline{\psi}(t; \gamma), \quad \tilde{\xi}(t; \gamma, 0) = \underline{\xi}(t; \gamma), \quad \tilde{\zeta}(t; \gamma, 0) = \underline{\zeta}(t; \gamma), \text{ and} \tag{6.14}$$

$$\tilde{\psi}(t; \gamma, 1,) = \overline{\psi}(t; \gamma), \quad \tilde{\xi}(t; \gamma, 1) = \overline{\xi}(t; \gamma), \quad \tilde{\zeta}(t; \gamma, 1) = \overline{\zeta}(t; \gamma). \tag{6.15}$$

Case II. In this case, we have considered Eq. (6.8) with initial conditions as

$$\psi(0) = 100, \ \xi(0) = 0, \ \text{and} \ \tilde{\zeta}(0) = (0.5, \ 1, \ 1.5) \tag{6.16}$$

using γ-cut; the triangular fuzzy initial conditions become

$$\psi(0) = 100, \ \xi(0) = 0, \ \text{and} \ \tilde{\zeta}(0; \gamma) = [0.5\gamma + 0.5, \ -0.5\gamma + 1.5] \tag{6.17}$$

by using the double parametric form; Eq. (6.17) reduces to

$$\psi(0) = 100, \ \xi(0) = 0, \ \text{and} \ \tilde{\zeta}(0; \gamma, \beta) = (-\gamma + 1)\beta + 0.5\gamma + 0.5 = \eta. \tag{6.18}$$

Implementing FRDTM to the Eqs. (6.8) and (6.18), we, respectively, obtain the Eq. (6.10) and the transformed fuzzy initial conditions as follows:

$$\tilde{\psi}_0(0) = 100, \ \tilde{\xi}_0(0) = 0, \ \text{and} \ \tilde{\zeta}_0(\gamma, \beta) = \eta. \tag{6.19}$$

Substituting Eq. (6.19) into Eq. (6.10), the following expressions are obtained successively for $k = 0, 1, 2, \ldots$

$$\tilde{\psi}_1(\gamma, \beta) = \frac{1}{\Gamma(1 + \alpha_1)} \left(s' - 100\mu - 100b\eta\right),$$

$$\tilde{\xi}_1(\gamma, \beta) = \frac{100b\eta}{\Gamma(1 + \alpha_2)},$$

$$\tilde{\zeta}_1(\gamma, \beta) = \frac{-r\eta}{\Gamma(1 + \alpha_3)},$$

$$\tilde{\psi}_2(\gamma, \beta) = \frac{\Gamma(1 + \alpha_1)}{\Gamma(1 + 2\alpha_1)}$$
$$\left[\left(-\frac{\mu(-100b\eta - 100\mu + s')}{\Gamma(1 + \alpha_1)} - b\left(\frac{-100r\eta}{\Gamma(1 + \alpha_3)} + \frac{(-100b\eta - 100\mu + s')\eta}{\Gamma(1 + \alpha_1)}\right)\right)\right],$$

$$\tilde{\xi}_2(\gamma, \beta) = \frac{\Gamma(1 + \alpha_2)}{\Gamma(1 + 2\alpha_2)}\left[b\left(\frac{-100r\eta}{\Gamma(1 + \alpha_3)} + \frac{(-100b\eta - 100\mu + s')\eta}{\Gamma(1 + \alpha_1)}\right) - \frac{100\varepsilon b\eta}{\Gamma(1 + \alpha_2)}\right],$$

$$\tilde{\zeta}_2(\gamma, \beta) = \frac{\Gamma(1 + \alpha_3)}{\Gamma(1 + 2\alpha_3)}\left[\frac{\eta r^2}{\Gamma(1 + \alpha_3)} + \frac{100cb\eta}{\Gamma(1 + \alpha_2)}\right],$$

$$\tilde{\psi}_3(\gamma,\beta) = \frac{\Gamma(1+2\alpha_1)}{\Gamma(1+3\alpha_1)}$$

$$\left(-\mu\left[\frac{\Gamma(1+\alpha_1)}{\Gamma(1+2\alpha_1)}\left(\left(\begin{array}{c}-\frac{\mu(-100b\eta-100\mu+s')}{\Gamma(1+\alpha_1)}\\-b\left(\frac{-100r\eta}{\Gamma(1+\alpha_3)}+\frac{(-100b\eta-100\mu+s')\eta}{\Gamma(1+\alpha_1)}\right)\end{array}\right)\right)\right)\right.$$

$$\left.\begin{array}{c}100\left(\frac{\Gamma(1+\alpha_3)}{\Gamma(1+2\alpha_3)}\left[\frac{r^2\eta}{\Gamma(1+\alpha_3)}+\frac{100b\eta c}{\Gamma(1+\alpha_2)}\right]\right)\\-\left(\frac{-r\eta}{\Gamma(1+\alpha_1)\Gamma(1+\alpha_3)}(s'-100\mu-100b\eta)\right)\\-b\left[-\left(\frac{\Gamma(1+\alpha_1)}{\Gamma(1+2\alpha_1)}\left[\left(\begin{array}{c}-\frac{\mu(-100b\eta-100\mu+s')}{\Gamma(1+\alpha_1)}\\-b\left(\frac{-100r\eta}{\Gamma(1+\alpha_3)}+\frac{(-100b\eta-100\mu+s')\eta}{\Gamma(1+\alpha_1)}\right)\end{array}\right)\right]\right)\right]\eta\right),$$

$$\tilde{\xi}_3(\gamma,\beta) = \frac{\Gamma(1+2\alpha_2)}{\Gamma(1+3\alpha_2)}$$

$$\left(-\varepsilon\frac{\Gamma(1+\alpha_2)}{\Gamma(1+2\alpha_2)}\left[b\left(\frac{-100r\eta}{\Gamma(1+\alpha_3)}+\frac{(-100b\eta-100\mu+s')\eta}{\Gamma(1+\alpha_1)}\right)-\frac{100b\eta\varepsilon}{\Gamma(1+\alpha_2)}\right]\right.$$

$$\left.+b\left(\begin{array}{c}100\left(\frac{\Gamma(1+\alpha_3)}{\Gamma(1+2\alpha_3)}\left[\frac{r^2\eta}{\Gamma(1+\alpha_3)}+\frac{100b\eta c}{\Gamma(1+\alpha_2)}\right]\right)-\left(\begin{array}{c}\frac{-r\eta}{\Gamma(1+\alpha_1)\Gamma(1+\alpha_3)}\\(s'-100\mu-100b\eta)\end{array}\right)\\-\left(\frac{\Gamma(1+\alpha_1)}{\Gamma(1+2\alpha_1)}\left[\left(\begin{array}{c}-\frac{\mu(-100b\eta-100\mu+s')}{\Gamma(1+\alpha_1)}\\-b\left(\frac{-100r\eta}{\Gamma(1+\alpha_3)}+\frac{(-100b\eta-100\mu+s')\eta}{\Gamma(1+\alpha_1)}\right)\end{array}\right)\right]\right)\end{array}\right)\eta\right),$$

$$\tilde{\zeta}_3(\gamma,\beta) = \frac{\Gamma(1+2\alpha_3)}{\Gamma(1+3\alpha_3)}\left(\begin{array}{c}c\frac{\Gamma(1+\alpha_2)}{\Gamma(1+2\alpha_2)}\left[b\left(\frac{-100r\eta}{\Gamma(1+\alpha_3)}+\frac{(-100b\eta-100\mu+s')\eta}{\Gamma(1+\alpha_1)}\right)-\frac{100b\eta\varepsilon}{\Gamma(1+\alpha_2)}\right]\\-r\frac{\Gamma(1+\alpha_3)}{\Gamma(1+2\alpha_3)}\left[\frac{r^2\eta}{\Gamma(1+\alpha_3)}+\frac{100cb\eta}{\Gamma(1+\alpha_2)}\right]\end{array}\right).$$

Similarly, repeating the process, we may have all the values of $\{\psi_k\}_{k=0}^{\infty}$, $\{\xi_k\}_{k=0}^{\infty}$, and $\{\zeta_k\}_{k=0}^{\infty}$. Taking inverse differential transform to $\{\psi_k\}_{k=0}^{\infty}$, $\{\xi_k\}_{k=0}^{\infty}$, and $\{\zeta_k\}_{k=0}^{\infty}$, the following nth-order approximate solutions are obtained:

$$\begin{cases}\tilde{\psi}_n(t,\gamma,\beta) = \sum_{k=0}^{n}\tilde{\psi}_k(\gamma,\beta)\,t^{\alpha_1 k},\\[2mm]\tilde{\xi}_n(t,\gamma,\beta) = \sum_{k=0}^{n}\tilde{\xi}_k(\gamma,\beta)\,t^{\alpha_2 k},\\[2mm]\tilde{\zeta}_n(t,\gamma,\beta) = \sum_{k=0}^{n}\tilde{\zeta}_k(\gamma,\beta)\,t^{\alpha_3 k}.\end{cases} \tag{6.20}$$

The lower- and upper-bound solutions of the model may be evaluated by substituting $\beta = 0$ and $\beta = 1$, respectively. Mathematically, we may write

$$\tilde{\psi}(t;\gamma,0,) = \underline{\psi}(t;\gamma),\quad \tilde{\xi}(t;\gamma,0) = \underline{\xi}(t;\gamma),\quad \tilde{\zeta}(t;\gamma,0) = \underline{\zeta}(t;\gamma), \tag{6.21}$$

and

$$\tilde{\psi}(t; \gamma, 1,) = \overline{\psi}(t; \gamma), \quad \tilde{\xi}(t; \gamma, 1) = \overline{\xi}(t; \gamma), \quad \tilde{\zeta}(t; \gamma, 1) = \overline{\zeta}(t; \gamma), \qquad (6.22)$$

respectively.

Case III. In this case, the authors have considered two initial conditions as the triangular fuzzy number, which are as follows:

$$\tilde{\psi}(0) = (99.5, 100, 100.5), \quad \tilde{\xi}(0) = 0, \text{ and } \tilde{\zeta}(0) = (0.5, 1, 1.5). \qquad (6.23)$$

Converting the Eq. (6.23) again into single parametric form, we obtain

$$\tilde{\psi}(0; \gamma) = (0.5\gamma + 99.5, -0.5\gamma + 100.5), \tilde{\xi}(0) = 0, \text{ and}$$
$$\tilde{\zeta}(0; \gamma) = [0.5\gamma + 0.5, -0.5\gamma + 1.5]. \qquad (6.24)$$

By the double parametric form, Eq. (6.24) reduces to

$$\tilde{\psi}(0; \gamma, \beta) = \beta(-\gamma + 1) + 0.5\gamma + 99.5 = \eta_1, \tilde{\xi}(0) = 0, \text{ and}$$
$$\tilde{\zeta}(0; \gamma, \beta) = (-\gamma + 1)\beta + 0.5\gamma + 0.5 = \eta_2. \qquad (6.25)$$

Employing FRDTM to Eqs. (6.8) and (6.25), we obtain Eq. (6.10) and the following reduced differential transformed fuzzy initial conditions:

$$\tilde{\psi}_0(\gamma, \beta) = \eta_1, \quad \tilde{\xi}_0(0) = 0, \quad \text{and} \quad \tilde{\zeta}_0(\gamma, \beta) = \eta_2. \qquad (6.26)$$

Using Eq. (6.26) into Eq. (6.10), the following expressions for $\{\psi_k\}_{k=0}^{\infty}$, $\{\xi_k\}_{k=0}^{\infty}$, and $\{\zeta_k\}_{k=0}^{\infty}$ are obtained successively:

$$\tilde{\psi}_1(\gamma, \beta) = \frac{1}{\Gamma(1+\alpha_1)}(s' - \mu\eta_1 - b\eta_1\eta_2),$$

$$\tilde{\xi}_1(\gamma, \beta) = \frac{b\eta_1\eta_2}{\Gamma(1+\alpha_2)},$$

$$\tilde{\zeta}_1(\gamma, \beta) = \frac{-r\eta_2}{\Gamma(1+\alpha_3)},$$

$$\tilde{\psi}_2(\gamma, \beta) = \frac{\Gamma(1+\alpha_1)}{\Gamma(1+2\alpha_1)}\left[\left(-\frac{\mu(-b\eta_1\eta_2 - \mu\eta_1 + s')}{\Gamma(1+\alpha_1)} - b\left(\begin{array}{c}\frac{-r\eta_2\eta_1}{\Gamma(1+\alpha_3)}\\ +\frac{(-b\eta_1\eta_2-\mu\eta_1+s')\eta_2}{\Gamma(1+\alpha_1)}\end{array}\right)\right)\right],$$

$$\tilde{\xi}_2(\gamma, \beta) = \frac{\Gamma(1+\alpha_2)}{\Gamma(1+2\alpha_2)}\left[b\left(\frac{-r\eta_2\eta_1}{\Gamma(1+\alpha_3)} + \frac{(-b\eta_1\eta_2-\mu\eta_1+s')\eta_2}{\Gamma(1+\alpha_1)}\right) - \frac{\varepsilon b\eta_1\eta_2}{\Gamma(1+\alpha_2)}\right],$$

$$\tilde{\xi}_2(\gamma, \beta) = \frac{\Gamma(1+\alpha_3)}{\Gamma(1+2\alpha_3)}\left[\frac{r^2\eta_2}{\Gamma(1+\alpha_3)} + \frac{b\eta_1\eta_2 c}{\Gamma(1+\alpha_2)}\right],$$

$$\tilde{\psi}_3(\gamma, \beta) = \frac{\Gamma(1+2\alpha_1)}{\Gamma(1+3\alpha_1)}$$

$$\left(\begin{array}{c} -\mu\left(\frac{\Gamma(1+\alpha_1)}{\Gamma(1+2\alpha_1)}\left(\left(\begin{array}{c}-\frac{\mu(-b\eta_1\eta_2-\mu\eta_1+s')}{\Gamma(1+\alpha_1)}\\ -b\left(\frac{-r\eta_2\eta_1}{\Gamma(1+\alpha_3)} + \frac{(-b\eta_1\eta_2-\mu\eta_1+s')\eta_2}{\Gamma(1+\alpha_1)}\right)\end{array}\right)\right)\right)\\ -b\left(\begin{array}{c}\eta_1\left(\frac{\Gamma(1+\alpha_3)}{\Gamma(1+2\alpha_3)}\left[\frac{r^2\eta_2}{\Gamma(1+\alpha_3)} + \frac{cb\eta_1\eta_2}{\Gamma(1+\alpha_2)}\right]\right)\\ -\left(\frac{-r\eta_2}{\Gamma(1+\alpha_1)\Gamma(1+\alpha_3)}(s'-\mu\eta_1-b\eta_1\eta_2)\right)\\ -\left(\frac{\Gamma(1+\alpha_1)}{\Gamma(1+2\alpha_1)}\left[\left(\begin{array}{c}-\frac{\mu(-b\eta_1\eta_2-\mu\eta_1+s')}{\Gamma(1+\alpha_1)}\\ -b\left(\frac{-r\eta_2\eta_1}{\Gamma(1+\alpha_3)} + \frac{(-b\eta_1\eta_2-\mu\eta_1+s')\eta_2}{\Gamma(1+\alpha_1)}\right)\end{array}\right)\right]\right)\eta_2\end{array}\right)\end{array}\right),$$

$$\tilde{\xi}_3(\gamma, \beta) = \frac{\Gamma(1+2\alpha_2)}{\Gamma(1+3\alpha_2)}$$

$$\left(\begin{array}{c} -\varepsilon\frac{\Gamma(1+\alpha_2)}{\Gamma(1+2\alpha_2)}\left[b\left(\frac{-r\eta_2\eta_1}{\Gamma(1+\alpha_3)} + \frac{(-b\eta_1\eta_2-\mu\eta_1+s')\eta_2}{\Gamma(1+\alpha_1)}\right) - \frac{\varepsilon b\eta_1\eta_2}{\Gamma(1+\alpha_2)}\right]\\ +b\left(\begin{array}{c}\eta_1\left(\frac{\Gamma(1+\alpha_3)}{\Gamma(1+2\alpha_3)}\left[\frac{r^2\eta_2}{\Gamma(1+\alpha_3)} + \frac{cb\eta_1\eta_2}{\Gamma(1+\alpha_2)}\right]\right) - \left(\begin{array}{c}\frac{-r\eta_2}{\Gamma(1+\alpha_1)\Gamma(1+\alpha_3)}\\ (s'-\mu\eta_1-b\eta_1\eta_2)\end{array}\right)\\ -\left(\frac{\Gamma(1+\alpha_1)}{\Gamma(1+2\alpha_1)}\left[\left(\begin{array}{c}-\frac{\mu(-b\eta_1\eta_2-\mu\eta_1+s')}{\Gamma(1+\alpha_1)}-\\ b\left(\frac{-r\eta_2\eta_1}{\Gamma(1+\alpha_3)} + \frac{(-b\eta_1\eta_2-\mu\eta_1+s')\eta_2}{\Gamma(1+\alpha_1)}\right)\end{array}\right)\right]\right)\eta_2\end{array}\right)\end{array}\right),$$

$$\tilde{\zeta}_3(\gamma, \beta) = \frac{\Gamma(1+2\alpha_3)}{\Gamma(1+3\alpha_3)}$$

$$\left(\begin{array}{c} c\frac{\Gamma(1+\alpha_2)}{\Gamma(1+2\alpha_2)}\left[\begin{array}{c}b\left(\frac{-r\eta_2\eta_1}{\Gamma(1+\alpha_3)} + \frac{(-b\eta_1\eta_2-\mu\eta_1+s')\eta_2}{\Gamma(1+\alpha_1)}\right)\\ -\frac{\varepsilon b\eta_1\eta_2}{\Gamma(1+\alpha_2)}\end{array}\right]\\ -r\frac{\Gamma(1+\alpha_3)}{\Gamma(1+2\alpha_3)}\left[\frac{r^2\eta_2}{\Gamma(1+\alpha_3)} + \frac{cb\eta_1\eta_2}{\Gamma(1+\alpha_2)}\right]\end{array}\right).$$

Similarly, the rest of the components can be evaluated. Now, applying the inverse differential transform (as in the previous cases) to $\{\psi_k\}_{k=0}^{\infty}$, $\{\xi_k\}_{k=0}^{\infty}$, and $\{\zeta_k\}_{k=0}^{\infty}$, we have the nth-order approximate solutions as (as in the previous cases):

$$
\begin{cases}
\tilde{\psi}_n (t \, ; \gamma, \beta) = \sum_{k=0}^{n} \tilde{\psi}_k (\gamma, \beta) \, t^{\alpha_1 k}, \\
\tilde{\xi}_n (t \, ; \gamma, \beta) = \sum_{k=0}^{n} \tilde{\xi}_k (\gamma, \beta) \, t^{\alpha_2 k}, \\
\tilde{\zeta}_n (t \, ; \gamma, \beta) = \sum_{k=0}^{n} \tilde{\zeta}_k (\gamma, \beta) \, t^{\alpha_3 k}.
\end{cases}
\tag{6.27}
$$

Equation (6.27) gives the exact solution of this model if we obtain the following expressions as

$$
\begin{cases}
\tilde{\psi} (t \, ; \gamma, \beta) = \lim_{n \to \infty} \tilde{\psi}_n (t \, ; \gamma, \beta), \\
\tilde{\xi} (t \, ; \gamma, \beta) = \lim_{n \to \infty} \tilde{\xi}_n (t \, ; \gamma, \beta), \\
\tilde{\zeta} (t \, ; \gamma, \beta) = \lim_{n \to \infty} \tilde{\zeta}_n (t \, ; \gamma, \beta).
\end{cases}
\tag{6.28}
$$

Substituting $\beta = 0$ and $\beta = 1$ in the solution of the model, we obtain the lower- and upper-bound solutions of the model. Mathematically, the following expressions are obtained, respectively:

$$
\tilde{\psi} (t \, ; \gamma, 0) = \underline{\psi} (t \, ; \gamma), \quad \tilde{\xi} (t \, ; \gamma, 0) = \underline{\xi} (t \, ; \gamma), \quad \tilde{\zeta} (t \, ; \gamma, 0) = \underline{\zeta} (t \, ; \gamma).
\tag{6.29}
$$

$$
\tilde{\psi} (t \, ; \gamma, 1) = \overline{\psi} (t \, ; \gamma), \quad \tilde{\xi} (t \, ; \gamma, 1) = \overline{\xi} (t \, ; \gamma), \quad \tilde{\zeta} (t \, ; \gamma, 1) = \overline{\zeta} (t \, ; \gamma).
\tag{6.30}
$$

Similarly, we may consider all the three initial conditions as fuzzy number in Case IV and may proceed as per the above procedure.

6.3   NUMERICAL RESULTS AND DISCUSSION

Various numerical simulations have been performed by taking different values of parameters involved in the titled problem and initial conditions. In order to obtain the approximate solution of this model, we have considered the values of the parameter as $s' = 0.272$ (day/mm^3), $\mu = 0.00136$ (day/mm^3), $b = 0.00027$ (day/virion/mm^3), $\varepsilon = 0.33$ (day/mm^3), $r = 2.0$ (day) and $c = 50$ (virion/CLM/day). Also, the initial conditions are given by $\psi (0) = 100, \xi (0) = 0$ and $\zeta (0) = 1$. The values of the parameters and the initial conditions are reported in the references [4–6]. All the calculations are computed by considering the third-order ($n = 3$) approximate solutions. The initial conditions of the title model are taken as TFN in three different cases. Tables 6.1–6.3, respectively, represent the lower- and upper-bound solution of the model for the Cases I–III at different values of γ and t. Also, Tables 6.1-6.3 display the comparison of the present crisp solutions with the existing solutions (third order) solved by Merdan and Khan [16]. From Tables 6.1–6.3, it is concluded that the crisp solution of the model lies in between lower and upper bounds of the fuzzy and interval solutions. Fuzzy and interval solutions of the title problem for particular Cases I–III are portrayed, respectively, in Figs. 6.1–6.3 and

Table 6.1: Fuzzy and crisp solutions of Case I at $\alpha_1 = \alpha_2 = \alpha_3 = 1$

$t \rightarrow$		0	0.3	0.6	0.9	1.0
$\gamma = 0$	$[\underline{\psi}, \overline{\psi}]$	[99.5, 100.5]	[99.53, 100.53]	[99.57, 100.56]	[99.60, 100.60]	[99.61, 100.61]
	$[\underline{\xi}, \overline{\xi}]$	[0,0]	[0.006, 0.006]	[0.0108, 0.0108]	[0.019, 0.019]	[0.023, 0.023]
	$[\underline{\zeta}, \overline{\zeta}]$	[1,1]	[0.578, 0.578]	[0.264, 0.264]	[-0.314, -0.316]	[-0.630, -0.633]
$\gamma = 0.3$	$[\underline{\psi}, \overline{\psi}]$	[99.65, 100.35]	[99.68, 100.38]	[99.72, 100.41]	[99.75, 100.45]	[99.76, 100.46]
	$[\underline{\xi}, \overline{\xi}]$	[0,0]	[0.006, 0.006]	[0.010, 0.010]	[0.018, 0.018]	[0.022, 0.023]
	$[\underline{\zeta}, \overline{\zeta}]$	[1,1]	[0.578, 0.578]	[0.264, 0.264]	[-0.315, -0.315]	[-0.631, -0.633]
$\gamma = 0.6$	$[\underline{\psi}, \overline{\psi}]$	[99.80, 100.20]	[99.83, 100.23]	[99.87, 100.26]	[99.90, 100.30]	[99.91, 100.31]
	$[\underline{\xi}, \overline{\xi}]$	[0,0]	[0.006, 0.006]	[0.010, 0.010]	[0.018, 0.018]	[0.0230, 0.023]
	$[\underline{\zeta}, \overline{\zeta}]$	[1,1]	[0.578, 0.578]	[0.264, 0.264]	[-0.315, -0.315]	[-0.632, -0.632]
$\gamma = 0.9$	$[\underline{\psi}, \overline{\psi}]$	[99.95, 100.05]	[99.98, 100.08]	[100.01, 100.11]	[100.05, 100.15]	[100.06, 100.16]
	$[\underline{\xi}, \overline{\xi}]$	[0,0]	[0.006, 0.006]	[0.010, 0.010]	[0.018, 0.018]	[0.023, 0.023]
	$[\underline{\zeta}, \overline{\zeta}]$	[1,1]	[0.578, 0.578]	[0.264, 0.264]	[-0.315, -0.315]	[-0.632, -0.632]
$\gamma = 1.0$	$[\underline{\psi}, \overline{\psi}]$	[100.0, 100.0]	[100.03, 100.03]	[100.06, 100.06]	[100.10, 100.10]	[100.11, 100.11]
	$[\underline{\xi}, \overline{\xi}]$	[0,0]	[0.006, 0.006]	[0.010, 0.010]	[0.0188, 0.0188]	[0.023, 0.0230]
	$[\underline{\zeta}, \overline{\zeta}]$	[1,1]	[0.578, 0.578]	[0.264, 0.264]	[-0.315, -0.315]	[-0.632, -0.632]
Merdan and Khan [16]						
	$[\underline{\psi}, \overline{\psi}]$	100.0	100.03	100.06	100.1	100.11
	$[\underline{\xi}, \overline{\xi}]$	0	0.006	0.010	0.0188	0.023
	$[\underline{\zeta}, \overline{\zeta}]$	1.0	0.578	0.264	-0.315	-0.6322

Table 6.2: Fuzzy and crisp solutions of Case II at $\alpha_1 = \alpha_2 = \alpha_3 = 1$

$t \rightarrow$		0	0.3	0.6	0.9	1.0
$\gamma = 0$	$[\underline{\psi}, \overline{\psi}]$	[100, 100]	[100.03, 100.03]	[100.07, 100.06]	[100.11, 100.09]	[100.12, 100.09]
	$[\underline{\xi}, \overline{\xi}]$	[0,0]	[0.003, 0.009]	[0.005, 0.016]	[0.009, 0.03]	[0.011, 0.034]
	$[\underline{\zeta}, \overline{\zeta}]$	[0.5, 1.5]	[0.289, 0.867]	[0.1323, 0.396]	[-0.157, -0.472]	[-0.316, -0.948]
$\gamma = 0.3$	$[\underline{\psi}, \overline{\psi}]$	[100, 100]	[100.03, 100.03]	[100.07, 100.06]	[100.10, 100.09]	[100.12, 100.10]
	$[\underline{\xi}, \overline{\xi}]$	[0,0]	[0.003, 0.008]	[0.007, 0.014]	[0.012, 0.025]	[0.014, 0.031]
	$[\underline{\zeta}, \overline{\zeta}]$	[0.65, 1.35]	[0.375, 0.780]	[0.172, 0.357]	[-0.204, -0.425]	[-0.411, -0.853]
$\gamma = 0.6$	$[\underline{\psi}, \overline{\psi}]$	[100, 100]	[100.03, 100.03]	[100.07, 100.06]	[100.10, 100.09]	[100.11, 100.10]
	$[\underline{\xi}, \overline{\xi}]$	[0, 0]	[0.004, 0.007]	[0.008, 0.012]	[0.015, 0.022]	[0.018, 0.027]
	$[\underline{\zeta}, \overline{\zeta}]$	[0.8, 1.2]	[0.462, 0.694]	[0.211, 0.317]	[-0.252, -0.378]	[-0.505, -0.758]
$\gamma = 0.9$	$[\underline{\psi}, \overline{\psi}]$	[100, 100]	[100.03, 100.03]	[100.07, 100.06]	[100.10, 100.10]	[100.11, 100.11]
	$[\underline{\xi}, \overline{\xi}]$	[0, 0]	[0.005, 0.006]	[0.010, 0.011]	[0.017, 0.019]	[0.021, 0.024]
	$[\underline{\zeta}, \overline{\zeta}]$	[0.95, 1.05]	[0.549, 0.607]	[0.251, 0.277]	[-0.299, -0.331]	[-0.600, -0.663]
$\gamma = 1.0$	$[\underline{\psi}, \overline{\psi}]$	[100, 100]	[100.03, 100.03]	[100.07, 100.07]	[100.10, 100.10]	[100.11, 100.11]
	$[\underline{\xi}, \overline{\xi}]$	[0, 0]	[0.006, 0.006]	[0.010, 0.010]	[0.018, 0.018]	[0.023, 0.023]
	$[\underline{\zeta}, \overline{\zeta}]$	[1, 1]	[0.578, 0.578]	[0.264, 0.264]	[-0.315, -0.315]	[-0.632, -0.632]
Merdan and Khan [16]						
	$[\underline{\psi}, \overline{\psi}]$	100	100.03	100.07	100.10	100.11
	$[\underline{\xi}, \overline{\xi}]$	0	0.006	0.010	0.018	0.023
	$[\underline{\zeta}, \overline{\zeta}]$	1	0.578	0.264	-0.315	-0.632

Table 6.3: Fuzzy and crisp solutions of Case III at $\alpha_1 = \alpha_2 = \alpha_3 = 1$

$t \rightarrow$		0	0.3	0.6	0.9	1.0
$\gamma = 0$	$[\underline{\psi}, \overline{\psi}]$	[99.5, 100.5]	[99.53, 100.53]	[99.57, 100.56]	[99.61, 100.59]	[99.62, 100.59]
	$[\underline{\xi}, \overline{\xi}]$	[0,0]	[0.002, 0.009]	[0.005, 0.016]	[0.009, 0.028]	[0.011, 0.034]
	$[\underline{\zeta}, \overline{\zeta}]$	[0.5, 1.5]	[0.289, 0.867]	[0.1323, 0.397]	[-0.157, -0.474]	[-0.315, -0.95]
$\gamma = 0.3$	$[\underline{\psi}, \overline{\psi}]$	[99.6, 100.3]	[99.68, 100.38]	[99.72, 100.41]	[99.75, 100.44]	[99.77, 100.45]
	$[\underline{\xi}, \overline{\xi}]$	[0,0]	[0.003, 0.008]	[0.007, 0.014]	[0.012, 0.025]	[0.014, 0.031]
	$[\underline{\zeta}, \overline{\zeta}]$	[0.65, 1.35]	[0.375, 0.781]	[0.171, 0.357]	[-0.204, -0.426]	[-0.410, -0.855]
$\gamma = 0.6$	$[\underline{\psi}, \overline{\psi}]$	[99.8, 100.2]	[99.83, 100.23]	[99.87, 100.26]	[99.90, 100.29]	[99.91, 100.30]
	$[\underline{\xi}, \overline{\xi}]$	[0, 0]	[0.004, 0.007]	[0.008, 0.013]	[0.015, 0.022]	[0.018, 0.027]
	$[\underline{\zeta}, \overline{\zeta}]$	[0.8, 1.2]	[0.462, 0.694]	[0.211, 0.317]	[-0.251, -0.378]	[-0.505, -0.759]
$\gamma = 0.9$	$[\underline{\psi}, \overline{\psi}]$	[99.95, 100.05]	[99.98, 100.08]	[100.02, 100.11]	[100.05, 100.15]	[100.6, 100.16]
	$[\underline{\xi}, \overline{\xi}]$	[0, 0]	[0.005, 0.006]	[0.010, 0.011]	[0.017, 0.019]	[0.021, 0.024]
	$[\underline{\zeta}, \overline{\zeta}]$	[0.95, 1.05]	[0.549, 0.607]	[0.251, 0.277]	[-0.299, -0.331]	[-0.600, -0.664]
$\gamma = 1.0$	$[\underline{\psi}, \overline{\psi}]$	[100, 100]	[100.03, 100.03]	[100.06, 100.06]	[100.1, 100.1]	[100.11, 100.11]
	$[\underline{\xi}, \overline{\xi}]$	[0, 0]	[0.006, 0.006]	[0.01, 0.01]	[0.018, 0.018]	[0.023, 0.023]
	$[\underline{\zeta}, \overline{\zeta}]$	[1, 1]	[0.578, 0.578]	[0.264, 0.264]	[-0.315, -0.315]	[-0.632, -0.632]
Merdan and Khan [16]						
	$[\underline{\psi}, \overline{\psi}]$	100	100.03	100.06	100.1	100.11
	$[\underline{\xi}, \overline{\xi}]$	0	0.006	0.01	0.018	0.023
	$[\underline{\zeta}, \overline{\zeta}]$	1	0.578	0.264	-0.315	-0.632

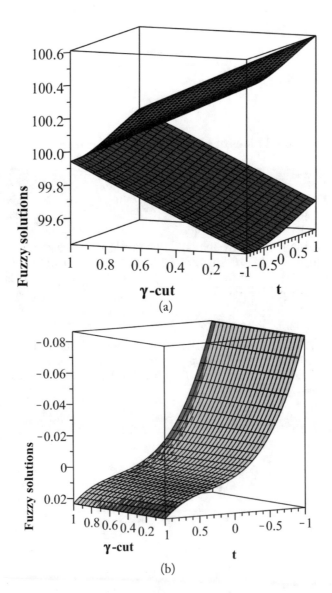

Figure 6.1: Fuzzy solution of Case I for (a) $\tilde{\psi}(t;\gamma)$ and (b) $\tilde{\xi}(t;\gamma)$ where $t \in [-1,-1]$ when $\alpha_1 = \alpha_2 = \alpha_3 = 1$. (*Continues.*)

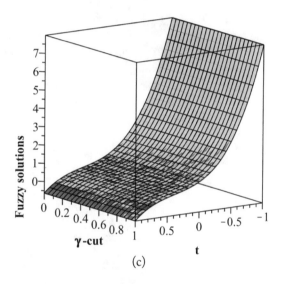

(c)

Figure 6.1: (*Continued.*) Fuzzy solution of Case I for (c) $\tilde{\zeta}(t;\gamma)$ where $t \in [-1, -1]$ when $\alpha_1 = \alpha_2 = \alpha_3 = 1$.

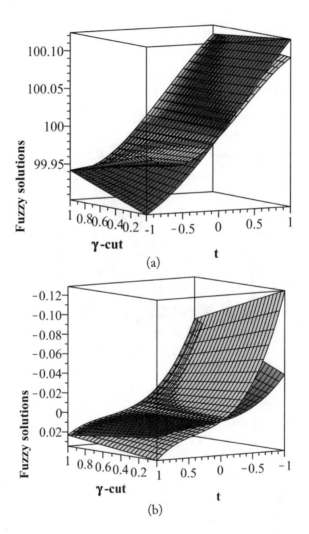

Figure 6.2: Fuzzy solution of Case II for (a) $\psi(t; \gamma)$ and (b) $\xi(t; \gamma)$ where $t \in [-1, -1]$ when $\alpha_1 = \alpha_2 = \alpha_3 = 1$. (*Continues.*)

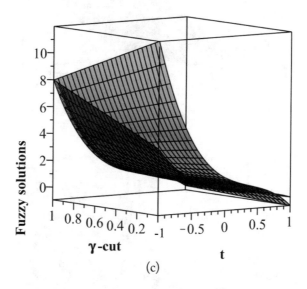

(c)

Figure 6.2: (*Continued.*) Fuzzy solution of Case II for (c) $\tilde{\xi}(t;\gamma)$ where $t \in [-1,-1]$ when $\alpha_1 = \alpha_2 = \alpha_3 = 1$.

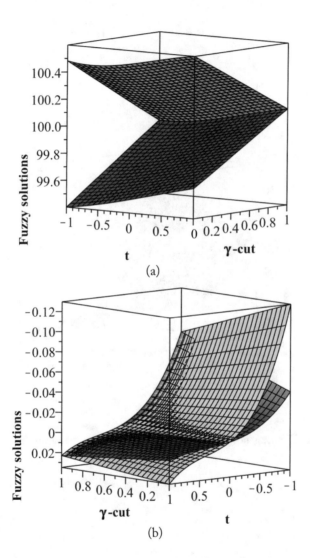

Figure 6.3: Fuzzy solution of Case III for (a) $\tilde{\psi}(t;\gamma)$ and (b) $\tilde{\xi}(t;\gamma)$ where $t \in [-1, -1]$ when $\alpha_1 = \alpha_2 = \alpha_3 = 1$. (*Continues.*)

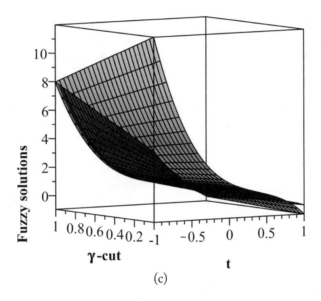

(c)

Figure 6.3: (*Continued.*) Fuzzy solution of Case III for (c) $\tilde{\zeta}(t;\gamma)$ where $t \in [-1,-1]$ when $\alpha_1 = \alpha_2 = \alpha_3 = 1$.

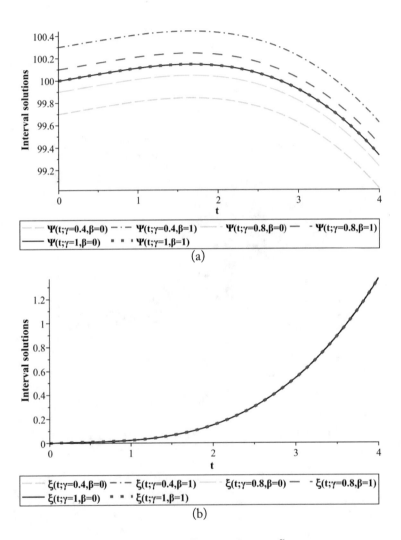

Figure 6.4: Interval solution of Case I for (a) $\tilde{\psi}\,(t;\gamma)$ and (b) $\tilde{\xi}\,(t;\gamma)$ where $\alpha_1 = \alpha_2 = \alpha_3 = 1$. (*Continues.*)

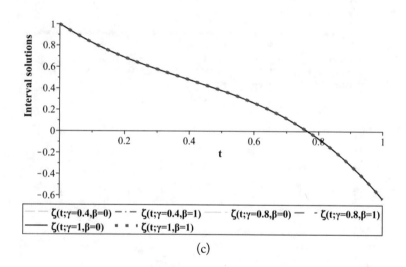

(c)

Figure 6.4: (*Continued.*) Interval solution of Case I for (c) $\tilde{\zeta}(t;\gamma)$ where $\alpha_1 = \alpha_2 = \alpha_3 = 1$.

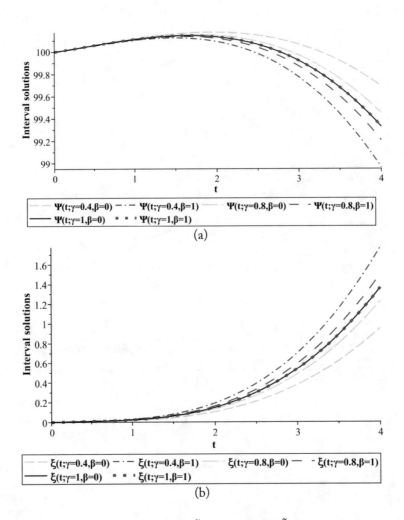

Figure 6.5: Interval solution of Case II for (a) $\tilde{\psi}\,(t;\gamma)$ and (b) $\tilde{\xi}\,(t;\gamma)$ where $\alpha_1 = \alpha_2 = \alpha_3 = 1$. (*Continues.*)

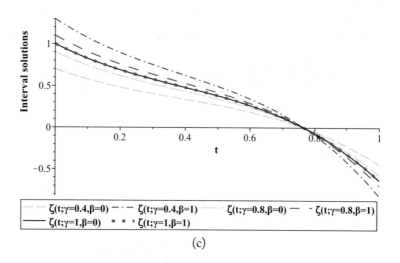

(c)

Figure 6.5: (*Continued.*) Interval solution of Case II for (c) $\tilde{\zeta}(t;\gamma)$ where $\alpha_1 = \alpha_2 = \alpha_3 = 1$.

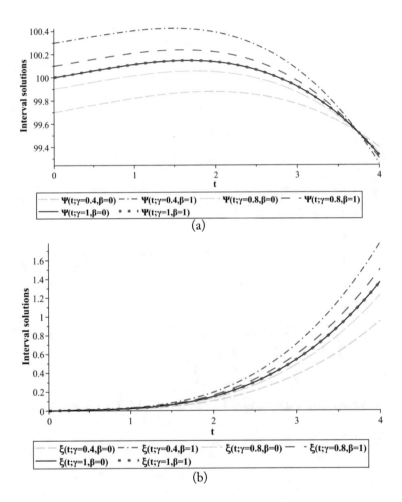

Figure 6.6: Interval solution of Case III for (a) $\tilde{\psi}\,(t;\gamma)$ and (b) $\tilde{\xi}\,(t;\gamma)$ where $\alpha_1 = \alpha_2 = \alpha_3 = 1$. (*Continues.*)

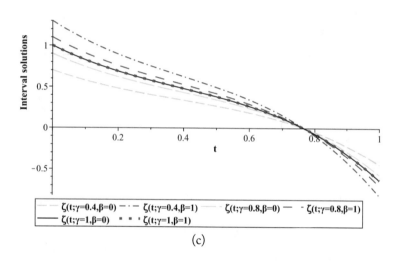

(c)

Figure 6.6: (*Continued.*) Interval solution of Case III for (c) $\tilde{\zeta}(t;\gamma)$ where $\alpha_1 = \alpha_2 = \alpha_3 = 1$.

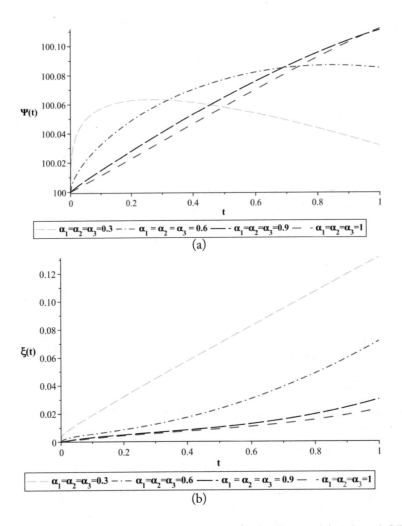

Figure 6.7: Crisp solution plots of (a) uninfected CD4+ T-cells and (b) infected CD4+ T-cells in plasm of titled problem at different values of α_1, α_2, and α_3. (*Continues.*)

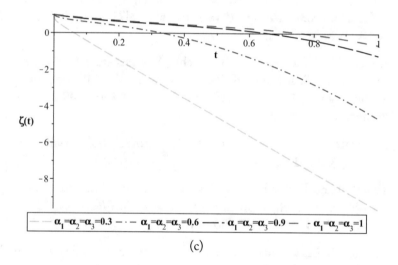

(c)

Figure 6.7: (*Continued.*) Crisp solution plots of (c) density of virion in plasm of titled problem at different values of α_1, α_2, and α_3.

Figs. 6.4–6.6. From these Figs. 6.4–6.6, one may notice that the crisp result ($\gamma = 1$) is the central line, and all the interval solutions are spread on both sides of the crisp results. It is worth mentioning that for all three cases, the present results at $\gamma = 1$ exactly match the third-order approximate solution of Merdan and Khan [16]. Figure 6.7 depicts the crisp solutions of HIV-1 CD4$^+$ T-cells at various values of α_1, α_2, and α_3.

6.4   REFERENCES

[1] R. Weiss. How does HIV cause AIDS?. *Science*, 260(5112):1273–1279, 1993. DOI: 10.1126/science.8493571. 75

[2] D. C. Douek, M. Roederer, and R. A. Koup. Emerging concepts in the immunopathogenesis of AIDS. *Annual Review of Medicine*, 60(1):471–484, 2009. DOI: 10.1146/annurev.med.60.041807.123549. 75

[3] R. W. Finberg, D. C. Diamond, D. B. Mitchel, Y. Rosenstein, G. Soman, T. C. Norman, S. L. Schreiber and S. J. Burakoff. Prevention of HIV-1 infection and preservation of CD4 function by the binding of CPFs to GP 120. *Science*, 249:287–291, 1990. DOI: 10.1126/science.2115689. 75

[4] H. C. Tuckwell and F. Y. M. Wan. On the behavior of solutions in viral dynamical models. *BioSystems*, 73(3):157–161, 2004. DOI: 10.1016/j.biosystems.2003.11.004. 84

[5] H. C. Tuckwell and F. Y. M. Wan. Nature of equilibria and effects of drug treatments in some simple viral population dynamical models. *IMA Journal of Mathematical Control and Information*, 17(4):311–327, 2000. DOI: 10.1093/imamci/17.4.311.

[6] B. H. Lichae, J. Biazar, and Z. Ayati. The fractional differential model of HIV-1 infection of CD4+ T-cells with description of the effect of antiviral drug treatment. *Computational and Mathematical Methods in Medicine*, 2019(4059549):12. DOI: 10.1155/2019/4059549. 76, 84

[7] A. Atangana and E. Alabaraoye. Solving a system of fractional partial differential equations arising in the model of HIV infection of CD4+ cells and attractor one-dimensional Keller–Segel equations. *Advances in Difference Equations*, pp. 94–107, 2013. DOI: 10.1186/1687-1847-2013-94.

[8] H. Buluta, D. Kumar, J. Singh, R. Swroop, and H. M. Baskonus. Analytic study for a fractional model of HIV infection of CD4+ T-lymphocyte cells. *Mathematics in Natural Science*, 2:33–43, 2018. DOI: 10.22436/mns.02.01.04. 76

[9] R. V. Culshaw and S. Ruan. A delay-differential equation model of HIV infection of CD4+ T-cells. *Mathematical Biosciences*, 165(1):27–39, 2000. DOI: 10.1016/s0025-5564(00)00006-7.

[10] M. Merdan, A. Gokdogan, and A. Yildirim. On the numerical solution of the model for HIV infection of CD4+ T-cells. *Computers and Mathematics with Applications*, 62(1):118–123, 2011. DOI: 10.1016/j.camwa.2011.04.058. 76

[11] S. Yuzbas. A numerical approach to solve the model for HIV infection of CD4+ T-cells. *Applied Mathematical Modelling*, 36(12):5876–5890, 2012. DOI: 10.1016/j.apm.2011.12.021.

[12] M. Ghoreishi, A. M. Ismail, and A. Alomari. Application of the homotopy analysis method for solving a model for HIV infection of CD4+ T-cells. *Mathematical and Computer Modelling*, 54(11–12):3007–3015, 2011. DOI: 10.1016/j.mcm.2011.07.029.

[13] M. Y. Ongun. Laplace Adomian decomposition method for solving a model for HIV infection of CD4+ T-cells. *Mathematical and Computer Modelling*, 53(5–6):597–603, 2011. DOI: doi.org/10.1016/j.mcm.2010.09.009.

[14] V. S. Erturk, Z. M. Odibat, and S. Momani. An approximate solution of a fractional-order differential equation model of human T-cell lymphotropic virus I (HTLV-I) infection of CD4+ T-cells. *Computers and Mathematics with Applications*, 62(3):996–1002, 2011. DOI: 10.1016/j.camwa.2011.03.091.

[15] A. A. M. Arafa, S. Rida, and M. Khalil. A fractional-order model of HIV infection with drug therapy effect. *Journal of the Egyptian Mathematical Society*, 22(3):538–543, 2014. DOI: 10.1016/j.joems.2013.11.001.

[16] M. Merdan and T. Khan. Homotopy perturbation method for solving viral dynamical model. *CU Fen-Edebiyat Fakultesi, Fen Bilimleri Dergisi*, 31:65–77, 2010. 84, 102

[17] A. A. M. Arafa, S. Z. Rida, and M. Khalil. The effect of antiviral drug treatment of human immunodeficiency virus type 1 (HIV-1) described by a fractional-order model. *Applied Mathematical Modelling*, 37(4):2189–2196, 2013. DOI: 10.1016/j.apm.2012.05.002.

[18] R. M. Jena, S. Chakraverty, and D. Baleanu. On the solution of imprecisely defined nonlinear time-fractional dynamical model of marriage. *Mathematics*, 7:689–704, 2019. DOI: 10.3390/math7080689. 75

[19] R. M. Jena, S. Chakraverty, and D. Baleanu. On new solutions of time-fractional wave equations arising in Shallow water wave propagation. *Mathematics*, 7:722–734, 2019. DOI: 10.3390/math7080722. 75

[20] R. M. Jena and S. Chakraverty. Solving time-fractional Navier–Stokes equations using homotopy perturbation Elzaki transform. *SN Applied Sciences*, 1(1):16, 2019. DOI: 10.1007/s42452-018-0016-9.

[21] R. M. Jena and S. Chakraverty. Residual power series method for solving time-fractional model of vibration equation of large membranes. *Journal of Applied and Computational Mechanics*, 5:603–615, 2019. DOI: 10.22055/jacm.2018.26668.1347.

[22] R. M. Jena and S. Chakraverty. A new iterative method based solution for fractional Black–Scholes option pricing equations (BSOPE). *SN Applied Sciences*, 1:95–105, 2019. DOI: 10.1007/s42452-018-0106-8.

[23] R. M. Jena and S. Chakraverty. Analytical solution of Bagley–Torvik equations using Sumudu transformation method. *SN Applied Sciences*, 1(3):246, 2019. DOI: 10.1007/s42452-019-0259-0.

[24] R. M. Jena, S. Chakraverty, and S. K. Jena. Dynamic response analysis of fractionally damped beams subjected to external loads using homotopy analysis method. *Journal of Applied and Computational Mechanics*, 5:355–366, 2019. DOI: 10.22055/JACM.2019.27592.1419 .

[25] S. Chakraverty, S. Tapaswini, and D. Behera. *Fuzzy Arbitrary Order System: Fuzzy Fractional Differential Equations and Applications*. John Wiley & Sons, 2016. DOI: 10.1002/9781119004233. 75

[26] S. Chakraverty, S. Tapaswini, and D. Behera. *Fuzzy Differential Equations and Applications for Engineers and Scientists*. Taylor & Francis Group, CRC Press, Boca Raton, FL, 2016. DOI: 10.1201/9781315372853.

[27] S. Chakraverty, D. M. Sahoo, and N. R. Mahato. *Concepts of Soft Computing: Fuzzy and ANN with Programming*. Springer, Singapore, 2019. DOI: 10.1007/978-981-13-7430-2. 75

CHAPTER 7

Time-Fractional Model of Hepatitis E Virus with Uncertain Parameters

7.1 INTRODUCTION

Hepatitis E is a biologically transmitted disorder resulting from the hepatitis E virus (HEV) [1]. Currently, HEV is considered as one of the main health problems globally. This virus causes infection of the liver, which mainly spread by the fecal-oral path. This results from drinking infected water, eating without washing hands, or poor sanitation. The first epidemic of HEV infection was recorded during 1955–1956 in New Delhi [2]. Another component is its possible relationship with malaria [3]. This contamination affected about 29,000 human beings because of fecal contamination of drinking water. According to the WHO, each year, about 20 million instances of HEV infections are recorded globally, and approximately 44,000 result in death in 2015 [4]. Most instances of infections are healed via the body's immune system within weeks. If not, HEV causes liver failure, which could cause loss of life for pregnant women, the elderly, or those who are sick. In this regard, mathematical models based on different infections are exhibited in [5–7]. In addition, a few mathematical models on HEV are given in the references [8–12]. Experimental facts have been utilized in [9] for pigs to gain knowledge about the viral dynamics of HEV. In [10], the authors studied the dynamics of hepatitis E and implemented strategies to sort out the difficulty that occurred in a Uganda displaced-persons camp. In [11], the authors proposed a mathematical model on HEV and provided a possible way to prevent HEV. A mathematical model regarding HEV infections in Shanghai, China, has been developed and studied in [12]. Differential equations are frequently used to explain a real-life phenomenon, but sometimes it may be difficult to observe the same when modeled with integer case. In this regard, fractional differential equations may be a better model to handle the titled problem. Various articles have been published in the literature where the fractional operator is implemented [5–8, 13–19].

The titled problem has been formulated using the Caputo fractional derivative. The purpose of this investigation is to analyze the dynamics of hepatitis E defined in a fractional sense with uncertainty. In this model, initial conditions are taken as uncertainty in terms of TFN.

7.2   MATHEMATICAL MODEL

It is an effective way to comprehend biological problems by introducing mathematical models and investigating their dynamical behaviors. In this analysis, we have considered the system of differential equations describing the behavior of HEV in the Caputo sense. To formulate the model for the HEV, the total population is divided into four sub-classes—susceptible individuals S, exposed individuals E, infected individuals I, and recovered individuals R—so $N = S + E + I + R$, where N is the total number of population. The environmental conditions play a vital role in the dynamics of HEV. So let P represent the viral load in the environment. The model based on the above sub-groups and their interaction with the environment is modeled using the following nonlinear system of differential equations:

$$\begin{cases} \frac{\partial S(t)}{\partial t} = \lambda \left(1 - \omega I(t)\right) - \left(bI(t) + \alpha_E P(t)\right) S(t) - v S(t) , \\[2mm] \frac{\partial E(t)}{\partial t} = \left(bI(t) + \alpha_E P(t)\right) S(t) - (v + \rho) E(t) + \lambda \omega I(t) , \\[2mm] \frac{\partial I(t)}{\partial t} = \rho E(t) - (v + \tau) I(t) , \\[2mm] \frac{\partial R(t)}{\partial t} = \tau I(t) - v R(t) , \\[2mm] \frac{\partial P(t)}{\partial t} = \delta I(t) - \sigma P(t) . \end{cases} \qquad (7.1)$$

The descriptions and values of the parameters are given in Table 7.1.

Equation (7.1) can be rewritten in the Caputo sense as:

$$\begin{cases} \frac{\partial^\alpha S(t)}{\partial t^\alpha} = \lambda \left(1 - \omega I(t)\right) - \left(bI(t) + \alpha_E P(t)\right) S(t) - v S(t) , \\[2mm] \frac{\partial^\alpha E(t)}{\partial t^\alpha} = \left(bI(t) + \alpha_E P(t)\right) S(t) - (v + \rho) E(t) + \lambda \omega I(t) , \\[2mm] \frac{\partial^\alpha I(t)}{\partial t^\alpha} = \rho E(t) - (v + \tau) I(t) , \\[2mm] \frac{\partial^\alpha R(t)}{\partial t^\alpha} = \tau I(t) - v R(t) , \\[2mm] \frac{\partial^\alpha P(t)}{\partial t^\alpha} = \delta I(t) - \sigma P(t), \end{cases} \qquad (7.2)$$

subject to the initial conditions [6]:

$$S(0) = 100, \quad E(0) = 35, \quad I(0) = 10, \quad R(0) = 1, \quad \text{and} \quad P(0) = 6. \qquad (7.3)$$

The main aim of this chapter is to obtain the solution of time-fractional HEV in Caputo sense using FRDTM in an uncertain environment.

Table 7.1: Descriptions of the parameters and their values involved in the model

Parameters	Descriptions	Values [6] (in day^{-1})
λ	Recruitment rate	0.8
b	Contact rate	0.0004
α_E	Rate of contact between S and P	0.0005
v	Natural human death rate	1/67.7
δ	Infections through environment	0.02
ρ	Rate of infection	0.02
τ	Recovery rate	0.023801429
σ	Virus decay in the environment	0.03
ω	Transfer rate of virus from mother to her child	0.02

7.3 MATHEMATICAL MODEL WITH FUZZY INITIAL CONDITIONS

Let us consider the following fuzzy time-fractional model of HEV:

$$
\begin{cases}
\frac{\partial^\alpha \tilde{S}(t)}{\partial t^\alpha} = \lambda \left(1 - \omega \, \tilde{I}(t)\right) - \left(b\tilde{I}(t) + \alpha_E \, \tilde{P}(t)\right) \tilde{S}(t) - v\tilde{S}(t), \\
\frac{\partial^\alpha \tilde{E}(t)}{\partial t^\alpha} = \left(b\tilde{I}(t) + \alpha_E \, \tilde{P}(t)\right) \tilde{S}(t) - (v + \rho) \, \tilde{E}(t) + \lambda \omega \tilde{I}(t), \\
\frac{\partial^\alpha \tilde{I}(t)}{\partial t^\alpha} = \rho \tilde{E}(t) - (v + \tau) \, \tilde{I}(t), \\
\frac{\partial^\alpha \tilde{R}(t)}{\partial t^\alpha} = \tau \, \tilde{I}(t) - v \, \tilde{R}(t), \\
\frac{\partial^\alpha \tilde{P}(t)}{\partial t^\alpha} = \delta \, \tilde{I}(t) - \sigma \, \tilde{P}(t).
\end{cases}
\tag{7.4}
$$

In this model, three initial conditions are considered as the TFN, which can be written as follows:

$$
\tilde{S}(0) = (95, 100, 105), \quad \tilde{E}(0) = (30, 35, 40),
$$
$$
\tilde{I}(0) = (5, 10, 15), \quad R(0) = 1, \quad \text{and} \quad P(0) = 6.
\tag{7.5}
$$

Using the γ-cut form, we can write Eq. (7.4) as

$$
\begin{cases}
\left[\frac{\partial^\alpha \underline{S}(t)}{\partial t^\alpha}, \frac{\partial^\alpha \overline{S}(t)}{\partial t^\alpha}\right] = \lambda\left(1 - \omega\left[\underline{I}(t), \overline{I}(t)\right]\right) - \left(b\left[\underline{I}(t), \overline{I}(t)\right]\right. \\
\qquad \left. + \alpha_E\left[\underline{P}(t), \overline{P}(t)\right]\right)\left[\underline{S}(t), \overline{S}(t)\right] - v\left[\underline{S}(t), \overline{S}(t)\right], \\
\left[\frac{\partial^\alpha \underline{E}(t)}{\partial t^\alpha}, \frac{\partial^\alpha \overline{E}(t)}{\partial t^\alpha}\right] = \left(b\left[\underline{I}(t), \overline{I}(t)\right] + \alpha_E\left[\underline{P}(t), \overline{P}(t)\right]\right)\left[\underline{S}(t), \overline{S}(t)\right] \\
\qquad - (v + \rho)\left[\underline{E}(t), \overline{E}(t)\right] + \lambda\omega\left[\underline{I}(t), \overline{I}(t)\right], \\
\left[\frac{\partial^\alpha \underline{I}(t)}{\partial t^\alpha}, \frac{\partial^\alpha \overline{I}(t)}{\partial t^\alpha}\right] = \rho\left[\underline{E}(t), \overline{E}(t)\right] - (v + \tau)\left[\underline{I}(t), \overline{I}(t)\right], \\
\left[\frac{\partial^\alpha \underline{R}(t)}{\partial t^\alpha}, \frac{\partial^\alpha \overline{R}(t)}{\partial t^\alpha}\right] = \tau\left[\underline{I}(t), \overline{I}(t)\right] - v\left[\underline{R}(t), \overline{R}(t)\right], \\
\left[\frac{\partial^\alpha \underline{P}(t)}{\partial t^\alpha}, \frac{\partial^\alpha \overline{P}(t)}{\partial t^\alpha}\right] = \delta\left[\underline{I}(t), \overline{I}(t)\right] - \sigma\left[\underline{P}(t), \overline{P}(t)\right],
\end{cases}
\tag{7.6}
$$

with fuzzy initial conditions:

$$
\begin{aligned}
\left[\underline{S}(0;\gamma), \overline{S}(0;\gamma)\right] &= [5\gamma + 95, -5\gamma + 105], \\
\left[\underline{E}(0;\gamma), \overline{E}(0;\gamma)\right] &= [5\gamma + 30, -5\gamma + 40], \\
\left[\underline{I}(0;\gamma), \overline{I}(0;\gamma)\right] &= [5\gamma + 5, -5\gamma + 15], \quad R(0) = 1, \ P(0) = 6,
\end{aligned}
\tag{7.7}
$$

where $\gamma \in [0, 1]$.

Next, by using the double parametric form (as discussed in previous chapters), Eqs. (7.6) and (7.7) can be expressed as

$$
\begin{cases}
\left\{\beta\left(\frac{\partial^\alpha \overline{S}(t\,;\gamma)}{\partial t^\alpha}-\frac{\partial^\alpha \underline{S}(t\,;\gamma)}{\partial t^\alpha}\right)+\frac{\partial^\alpha \underline{S}(t\,;\gamma)}{\partial t^\alpha}\right\}=\lambda\left(1-\omega\left(\beta\left(\overline{I}\left(t\,;\gamma\right)-\underline{I}\left(t\,;\gamma\right)\right)+\underline{I}\left(t\,;\gamma\right)\right)\right)\\
-\left(b\left(\beta\left(\overline{I}\left(t\,;\gamma\right)-\underline{I}\left(t\,;\gamma\right)\right)+\underline{I}\left(t\,;\gamma\right)\right)+\alpha_E\left(\beta\left(\overline{P}\left(t\,;\gamma\right)-\underline{P}\left(t\,;\gamma\right)\right)+\underline{P}\left(t\,;\gamma\right)\right)\right)\\
\left(\beta\left(\overline{S}\left(t\,;\gamma\right)-\underline{S}\left(t\,;\gamma\right)\right)+\underline{S}\left(t\,;\gamma\right)\right)-v\left(\beta\left(\overline{S}\left(t\,;\gamma\right)-\underline{S}\left(t\,;\gamma\right)\right)+\underline{S}\left(t\,;\gamma\right)\right),\\[2mm]
\left\{\beta\left(\frac{\partial^\alpha \overline{E}(t\,;\gamma)}{\partial t^\alpha}-\frac{\partial^\alpha \underline{E}(t\,;\gamma)}{\partial t^\alpha}\right)+\frac{\partial^\alpha \underline{E}(t\,;\gamma)}{\partial t^\alpha}\right\}=\begin{pmatrix}b\left(\beta\left(\overline{I}\left(t\,;\gamma\right)-\underline{I}\left(t\,;\gamma\right)\right)+\underline{I}\left(t\,;\gamma\right)\right)\\+\alpha_E\left(\beta\left(\overline{P}\left(t\,;\gamma\right)-\underline{P}\left(t\,;\gamma\right)\right)+\underline{P}\left(t\,;\gamma\right)\right)\end{pmatrix}\\
\left(\beta\left(\overline{S}\left(t\,;\gamma\right)-\underline{S}\left(t\,;\gamma\right)\right)+\underline{S}\left(t\,;\gamma\right)\right)-(v+\rho)\left(\beta\left(\overline{E}\left(t\,;\gamma\right)-\underline{E}\left(t\,;\gamma\right)\right)+\underline{E}\left(t\,;\gamma\right)\right)\\
+\lambda\omega\left(\beta\left(\overline{I}\left(t\,;\gamma\right)-\underline{I}\left(t\,;\gamma\right)\right)+\underline{I}\left(t\,;\gamma\right)\right),\\[2mm]
\left\{\beta\left(\frac{\partial^\alpha \overline{I}(t\,;\gamma)}{\partial t^\alpha}-\frac{\partial^\alpha \underline{I}(t\,;\gamma)}{\partial t^\alpha}\right)+\frac{\partial^\alpha \underline{I}(t\,;\gamma)}{\partial t^\alpha}\right\}=\rho\left(\beta\left(\overline{E}\left(t\,;\gamma\right)-\underline{E}\left(t\,;\gamma\right)\right)+\underline{E}\left(t\,;\gamma\right)\right)-(v+\tau)\\
\left(\beta\left(\overline{I}\left(t\,;\gamma\right)-\underline{I}\left(t\,;\gamma\right)\right)+\underline{I}\left(t\,;\gamma\right)\right),\\[2mm]
\left\{\beta\left(\frac{\partial^\alpha \overline{R}(t\,;\gamma)}{\partial t^\alpha}-\frac{\partial^\alpha \underline{R}(t\,;\gamma)}{\partial t^\alpha}\right)+\frac{\partial^\alpha \underline{R}(t\,;\gamma)}{\partial t^\alpha}\right\}=\tau\left(\beta\left(\overline{I}\left(t\,;\gamma\right)-\underline{I}\left(t\,;\gamma\right)\right)+\underline{I}\left(t\,;\gamma\right)\right)\\
-v\left(\beta\left(\overline{R}\left(t\,;\gamma\right)-\underline{R}\left(t\,;\gamma\right)\right)+\underline{R}\left(t\,;\gamma\right)\right),\\[2mm]
\left\{\beta\left(\frac{\partial^\alpha \overline{P}(t\,;\gamma)}{\partial t^\alpha}-\frac{\partial^\alpha \underline{P}(t\,;\gamma)}{\partial t^\alpha}\right)+\frac{\partial^\alpha \underline{P}(t\,;\gamma)}{\partial t^\alpha}\right\}=\delta\left(\beta\left(\overline{I}\left(t\,;\gamma\right)-\underline{I}\left(t\,;\gamma\right)\right)+\underline{I}\left(t\,;\gamma\right)\right)\\
-\sigma\left(\beta\left(\overline{P}\left(t\,;\gamma\right)-\underline{P}\left(t\,;\gamma\right)\right)+\underline{P}\left(t\,;\gamma\right)\right),
\end{cases}
$$

(7.8)

with initial conditions:

$$
\begin{cases}
\beta\left(\overline{S}\left(0\,;\gamma\right)-\underline{S}\left(0\,;\gamma\right)\right)+\underline{S}\left(0\,;\gamma\right)=\beta\left(-10\gamma+10\right)+5\gamma+95,\\
\beta\left(\overline{E}\left(0\,;\gamma\right)-\underline{E}\left(0\,;\gamma\right)\right)+\underline{E}\left(0\,;\gamma\right)=\beta\left(-10\gamma+10\right)+5\gamma+30,\\
\beta\left(\overline{I}\left(0\,;\gamma\right)-\underline{I}\left(0\,;\gamma\right)\right)+\underline{I}\left(0\,;\gamma\right)=\beta\left(-10\gamma+10\right)+5\gamma+5,\\
R\left(0\right)=1,\ P\left(0\right)=6.
\end{cases}
$$

(7.9)

Let us assume

$$
\beta\left(\frac{\partial^\alpha \overline{S}\left(t\,;\gamma\right)}{\partial t^\alpha}-\frac{\partial^\alpha \underline{S}\left(t\,;\gamma\right)}{\partial t^\alpha}\right)+\frac{\partial^\alpha \underline{S}\left(t\,;\gamma\right)}{\partial t^\alpha}=\frac{\partial^\alpha \tilde{S}\left(t\,;\gamma,\beta\right)}{\partial t^\alpha},
$$

$$
\beta\left(\frac{\partial^\alpha \overline{E}\left(t\,;\gamma\right)}{\partial t^\alpha}-\frac{\partial^\alpha \underline{E}\left(t\,;\gamma\right)}{\partial t^\alpha}\right)+\frac{\partial^\alpha \underline{E}\left(t\,;\gamma\right)}{\partial t^\alpha}=\frac{\partial^\alpha \tilde{E}\left(t\,;\gamma,\beta\right)}{\partial t^\alpha},
$$

$$\beta \left(\frac{\partial^\alpha \overline{I}\,(t\,;\gamma)}{\partial t^\alpha} - \frac{\partial^\alpha \underline{I}\,(t\,;\gamma)}{\partial t^\alpha} \right) + \frac{\partial^\alpha \underline{I}\,(t\,;\gamma)}{\partial t^\alpha} = \frac{\partial^\alpha \tilde{I}\,(t\,;\gamma,\beta)}{\partial t^\alpha},$$

$$\beta \left(\frac{\partial^\alpha \overline{R}\,(t\,;\gamma)}{\partial t^\alpha} - \frac{\partial^\alpha \underline{R}\,(t\,;\gamma)}{\partial t^\alpha} \right) + \frac{\partial^\alpha \underline{R}\,(t\,;\gamma)}{\partial t^\alpha} = \frac{\partial^\alpha \tilde{R}\,(t\,;\gamma,\beta)}{\partial t^\alpha},$$

$$\beta \left(\frac{\partial^\alpha \overline{P}\,(t\,;\gamma)}{\partial t^\alpha} - \frac{\partial^\alpha \underline{P}\,(t\,;\gamma)}{\partial t^\alpha} \right) + \frac{\partial^\alpha \underline{P}\,(t\,;\gamma)}{\partial t^\alpha} = \frac{\partial^\alpha \tilde{P}\,(t\,;\gamma,\beta)}{\partial t^\alpha},$$

$$\beta \left(\overline{S}\,(t\,;\gamma) - \underline{S}\,(t\,;\gamma) \right) + \underline{S}\,(t\,;\gamma) = \tilde{S}\,(t\,;\gamma,\beta),$$
$$\beta \left(\overline{E}\,(t\,;\gamma) - \underline{E}\,(t\,;\gamma) \right) + \underline{E}\,(t\,;\gamma) = \tilde{E}\,(t\,;\gamma,\beta),$$
$$\beta \left(\overline{I}\,(t\,;\gamma) - \underline{I}\,(t\,;\gamma) \right) + \underline{I}\,(t\,;\gamma) = \tilde{I}\,(t\,;\gamma,\beta),$$
$$\beta \left(\overline{R}\,(t\,;\gamma) - \underline{R}\,(t\,;\gamma) \right) + \underline{R}\,(t\,;\gamma) = \tilde{R}\,(t\,;\gamma,\beta),$$
$$\beta \left(\overline{P}\,(t\,;\gamma) - \underline{P}\,(t\,;\gamma) \right) + \underline{P}\,(t\,;\gamma) = \tilde{P}\,(t\,;\gamma,\beta),$$
$$\beta \left(\overline{S}\,(0\,;\gamma) - \underline{S}\,(0\,;\gamma) \right) + \underline{S}\,(0\,;\gamma) = \tilde{S}\,(0\,;\gamma,\beta),$$
$$\beta \left(\overline{E}\,(0\,;\gamma) - \underline{E}\,(0\,;\gamma) \right) + \underline{E}\,(0\,;\gamma) = \tilde{E}\,(0\,;\gamma,\beta),$$
$$\beta \left(\overline{I}\,(0\,;\gamma) - \underline{I}\,(0\,;\gamma) \right) + \underline{I}\,(0\,;\gamma) = \tilde{I}\,(0\,;\gamma,\beta),$$
$$\beta \left(-10\gamma + 10 \right) + 5\gamma + 95 = \eta_1, \quad \beta \left(-10\gamma + 10 \right) + 5\gamma + 30 = \eta_2,$$
$$\beta \left(-10\gamma + 10 \right) + 5\gamma + 5 = \eta_3.$$

Putting all the above equations in Eqs. (7.8) and (7.9), we will get the following expressions:

$$\begin{cases} \frac{\partial^\alpha \tilde{S}(t;\gamma,\beta)}{\partial t^\alpha} = \lambda \left(1 - \omega \tilde{I}\,(t;\gamma,\beta) \right) - \left(b\tilde{I}\,(t;\gamma,\beta) + \alpha_E \tilde{P}\,(t;\gamma,\beta) \right) \tilde{S}\,(t;\gamma,\beta) - v\tilde{S}\,(t;\gamma,\beta), \\[2mm] \frac{\partial^\alpha \tilde{E}(t;\gamma,\beta)}{\partial t^\alpha} = \left(b\tilde{I}\,(t;\gamma,\beta) + \alpha_E \tilde{P}\,(t;\gamma,\beta) \right) \tilde{S}\,(t;\gamma,\beta) - (v + \rho)\,\tilde{E}\,(t;\gamma,\beta) + \lambda \omega \tilde{I}\,(t;\gamma,\beta), \\[2mm] \frac{\partial^\alpha \tilde{I}(t;\gamma,\beta)}{\partial t^\alpha} = \rho \tilde{E}\,(t;\gamma,\beta) - (v + \tau)\,\tilde{I}\,(t;\gamma,\beta), \\[2mm] \frac{\partial^\alpha \tilde{R}(t;\gamma,\beta)}{\partial t^\alpha} = \tau \tilde{I}\,(t;\gamma,\beta) - v\tilde{R}\,(t;\gamma,\beta), \\[2mm] \frac{\partial^\alpha \tilde{P}(t;\gamma,\beta)}{\partial t^\alpha} = \delta \tilde{I}\,(t;\gamma,\beta) - \sigma \tilde{P}\,(t;\gamma,\beta), \end{cases}$$

(7.10)

with fuzzy initial conditions:

$$\tilde{S}\,(0\,;\gamma,\beta) = \eta_1, \quad \tilde{E}\,(0\,;\gamma,\beta) = \eta_2, \quad \tilde{I}\,(0\,;\gamma,\beta) = \eta_3, \quad R\,(0) = 1, \quad P\,(0) = 6. \quad (7.11)$$

Now, by using FRDTM to Eqs. (7.10), and (7.11) (fuzzy initial conditions), we obtain the following expressions, respectively:

$$
\begin{cases}
\tilde{S}_{k+1}(\gamma, \beta) = \frac{\Gamma(1+\alpha k)}{\Gamma(1+\alpha k+\alpha)} \left\{ \begin{array}{l} \lambda del(k) - \lambda\omega \tilde{I}_k(\gamma, \beta) - b\sum_{i=0}^{k} \tilde{I}_i(\gamma, \beta)\tilde{S}_{k-i}(\gamma, \beta) \\ -\alpha_E \sum_{i=0}^{k} \tilde{P}_i(\gamma, \beta)\tilde{S}_{k-i}(\gamma, \beta) - v\tilde{S}_k(\gamma, \beta) \end{array} \right\}, \\[12pt]
\tilde{E}_{k+1}(\gamma, \beta) = \frac{\Gamma(1+\alpha k)}{\Gamma(1+\alpha k+\alpha)} \left\{ \begin{array}{l} b\sum_{i=0}^{k} \tilde{I}_i(\gamma, \beta)\tilde{S}_{k-i}(\gamma, \beta) + \alpha_E \sum_{i=0}^{k} \tilde{P}_i(\gamma, \beta)\tilde{S}_{k-i}(\gamma, \beta) \\ -(v+\rho)\tilde{E}_k(\gamma, \beta) + \lambda\omega \tilde{I}_k(\gamma, \beta) \end{array} \right\}, \\[12pt]
\tilde{I}_{k+1}(\gamma, \beta) = \frac{\Gamma(1+\alpha k)}{\Gamma(1+\alpha k+\alpha)} \left\{ \rho\tilde{E}_k(\gamma, \beta) - (v+\tau)\tilde{I}_k(\gamma, \beta) \right\}, \\[6pt]
\tilde{R}_{k+1}(\gamma, \beta) = \frac{\Gamma(1+\alpha k)}{\Gamma(1+\alpha k+\alpha)} \left\{ \tau \tilde{I}_k(\gamma, \beta) - v\tilde{R}_k(\gamma, \beta) \right\}, \\[6pt]
\tilde{P}_{k+1}(\gamma, \beta) = \frac{\Gamma(1+\alpha k)}{\Gamma(1+\alpha k+\alpha)} \left\{ \delta \tilde{I}_k(\gamma, \beta) - \sigma \tilde{P}_k(\gamma, \beta) \right\},
\end{cases}
\tag{7.12}
$$

with transformed fuzzy initial conditions:

$$
\tilde{S}_0(\gamma, \beta) = \eta_1, \ \tilde{E}_0(\gamma, \beta) = \eta_2, \ \tilde{I}_0(\gamma, \beta) = \eta_3, \ \tilde{R}_0(\gamma, \beta) = 1, \ \tilde{P}_0(\gamma, \beta) = 6, \tag{7.13}
$$

where,

$$
del(k) = \begin{cases} 1, & k = 0, \\ 0, & k \neq 0. \end{cases} \tag{7.14}
$$

Substituting Eqs. (7.13) and (7.14) into Eq. (7.12) for $k = 0, 1, 2, \ldots$, the following expressions are obtained successively:

$$
\tilde{S}_1(\gamma, \beta) = \frac{-0.016\eta_3 + 0.8 - 0.0004\eta_3\eta_1 - 0.0177710487\eta_1}{\Gamma(1+\alpha)},
$$

$$
\tilde{E}_1(\gamma, \beta) = \frac{0.0004\eta_3\eta_1 + 0.003\eta_1 - 0.0347710487\eta_2 + 0.016\eta_3}{\Gamma(1+\alpha)},
$$

$$
\tilde{I}_1(\gamma, \beta) = \frac{0.02\eta_2 - 0.03857247774\eta_3}{\Gamma(1+\alpha)},
$$

$$
\tilde{R}_1(\gamma, \beta) = \frac{0.023801429\eta_3 - 0.01477104874}{\Gamma(1+\alpha)},
$$

$$
\tilde{P}_1(\gamma, \beta) = \frac{0.02\eta_3 - 0.18}{\Gamma(1+\alpha)}.
$$

In this manner, we can obtain all the values of $\{\tilde{S}_k\}_{k=0}^{\infty}$, $\{\tilde{E}_k\}_{k=0}^{\infty}$, $\{\tilde{I}_k\}_{k=0}^{\infty}$, $\{\tilde{R}_k\}_{k=0}^{\infty}$, and $\{\tilde{P}_k\}_{k=0}^{\infty}$. Using inverse differential transform to $\{\tilde{S}_k\}_{k=0}^{\infty}$, $\{\tilde{E}_k\}_{k=0}^{\infty}$, $\{\tilde{I}_k\}_{k=0}^{\infty}$, $\{\tilde{R}_k\}_{k=0}^{\infty}$,

$\{\tilde{P}_k\}_{k=0}^{\infty}$, and substituting the values of η_1, η_2, and η_3, we have the following nth order approximate solutions:

$$
\begin{cases}
\tilde{S}_n\,(t\,;\,\gamma,\beta) = \sum_{k=0}^{n} \tilde{S}_k\,(\gamma,\beta)\,t^{\alpha k}, \quad \tilde{E}_n\,(t\,;\,\gamma,\beta) = \sum_{k=0}^{n} \tilde{E}_k\,(\gamma,\beta)\,t^{\alpha k}, \\[2mm]
\tilde{I}_n\,(t\,;\,\gamma,\beta) = \sum_{k=0}^{n} \tilde{I}_k\,(\gamma,\beta)\,t^{\alpha k}, \quad \tilde{R}_n\,(t\,;\,\gamma,\beta) = \sum_{k=0}^{n} \tilde{R}_k\,(\gamma,\beta)\,t^{\alpha k}, \\[2mm]
\tilde{P}_n\,(t\,;\,\gamma,\beta) = \sum_{k=0}^{n} \tilde{P}_k\,(\gamma,\beta)\,t^{\alpha k},
\end{cases}
\tag{7.15}
$$

we may find the exact solution of the model if the series Eq. (7.15) converges to a particular solution. Similarly, one may consider other involved parameters, as well as initial conditions as fuzzy number, and the methodology may be similar as per the above procedure.

7.4   NUMERICAL RESULTS AND DISCUSSION

In this section, we have discussed the numerical results for the time-fractional model of HEV with uncertain parameters. The numerical results are computed by using FRDTM. All the numerical computations have been performed by considering various values of the parameters which are given in Table 7.1. All the numerical calculations and plots are presented by considering the third-order ($n = 3$) approximate solutions. Susceptible individuals (S), exposed individuals (E), and infected individuals (I) at $t = 0$ are considered to be the triangular fuzzy numbers. The lower- and upper-bound fuzzy solutions of the fractional HEV model are depicted in Fig. 7.1 by varying t from 0–30 at $\alpha = 1$. Figure 7.2 displays the effect of γ-cut on different classes individuals. One may conclude from Fig. 7.2 that the crisp solution ($\gamma = 1$) is the middle line and all other interval solutions are spread on both sides of the crisp results. Figure 7.3 presents the behavior of susceptible individuals (S), exposed individuals (E), infected individuals (I), recovered individuals (R), and density of viral load in the environment (P) at the particular values of fractional order that is $\alpha = 0.2, 0.3, 0.5, 0.7,$ and 0.9. Table 7.2 includes the lower- and upper-bounds interval solutions of the model at different values of t and γ. From Table 7.2, it is clear that the lower- and the upper-bound solutions are equal when $\gamma = 1$. From Figs. (7.3)a and (7.3)b, we can observe that as time increases, the number of $S(t)$ and $E(t)$ decreases. Further, the number of exposed individuals $I(t)$, recovered individuals $R(t)$, and density of the viral load $P(t)$ increases with an increase in time. From this observation, it is clear that the discussed model significantly depends on the fractional-order derivatives, which may help to explore the biological behavior of the proposed model.

Table 7.2: Lower- and upper-bound solutions of the model at $\alpha = 1$

$t \rightarrow$		0.4	0.8	1.0
$\gamma = 0.4$	$[\underline{S}, \overline{S}]$	[96.1590, 101.9736]	[95.3221, 100.9537]	[94.9052, 100.4461]
	$[\underline{E}, \overline{E}]$	[31.8264, 37.8930]	[31.6562, 37.7872]	[31.5722, 37.7346]
	$[\underline{I}, \overline{I}]$	[7.1461, 13.1022]	[7.2887, 13.2019]	[7.3586, 13.2510]
	$[\underline{R}, \overline{R}]$	[1.0612, 1.1179]	[1.1235, 1.2362]	[1.1550, 1.2954]
	$[\underline{P}, \overline{P}]$	[5.9846, 6.0321]	[5.9706, 6.0648]	[5.9641, 6.0227]
$\gamma = 0.8$	$[\underline{S}, \overline{S}]$	[98.0984, 100.0366]	[97.2018, 99.0789]	[96.7552, 98.6022]
	$[\underline{E}, \overline{E}]$	[33.8474, 35.8696]	[33.6974, 35.7410]	[33.6233, 35.6774]
	$[\underline{I}, \overline{I}]$	[9.1315, 11.11685]	[9.2598, 11.2308]	[9.3227, 11.2868]
	$[\underline{R}, \overline{R}]$	[1.0801, 1.0990]	[1.1610, 1.1986]	[1.2018, 1.2486]
	$[\underline{P}, \overline{P}]$	[6.0005, 6.0163]	[6.0020, 6.0334]	[6.0032, 6.0422]
$\gamma = 1.0$	$[\underline{S}, \overline{S}]$	[99.0677, 99.0677]	[98.1407, 98.1407]	[97.6791, 97.6791]
	$[\underline{E}, \overline{E}]$	[34.8583, 34.8583]	[34.7189, 34.7189]	[34.6500, 34.6500]
	$[\underline{I}, \overline{I}]$	[10.1241, 10.1241]	[10.2453, 10.2453]	[10.3048, 10.3048]
	$[\underline{R}, \overline{R}]$	[1.0896, 1.0896]	[1.1798, 1.1798]	[1.2252, 1.2252]
	$[\underline{P}, \overline{P}]$	[6.0084, 6.0084]	[6.0177, 6.0177]	[6.0227, 6.0227]

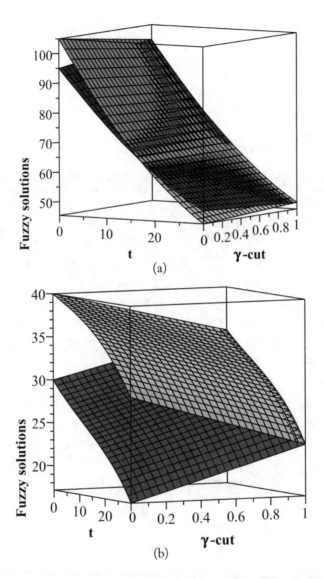

Figure 7.1: Fuzzy solution of fractional order HEV model for (a) $\tilde{S}(t;\gamma)$ and (b) $\bar{E}(t;\gamma)$ where $t \in [0, 30]$ and $\gamma \in [0, 1]$ when $\alpha = 1$. (*Continues.*)

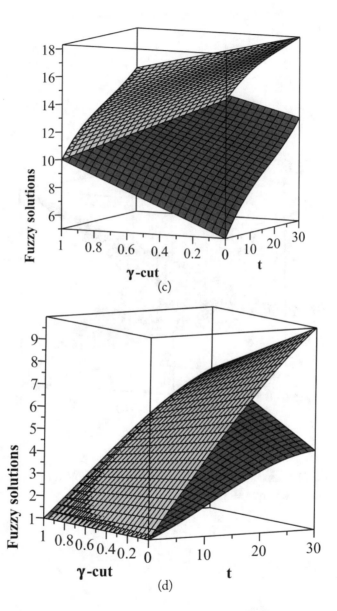

Figure 7.1: (*Continued.*) Fuzzy solution of fractional order HEV model for (c) $\tilde{I}(t;\gamma)$ and (d) $\tilde{R}(t;\gamma)$ where $t \in [0, 30]$ and $\gamma \in [0, 1]$ when $\alpha = 1$. (*Continues.*)

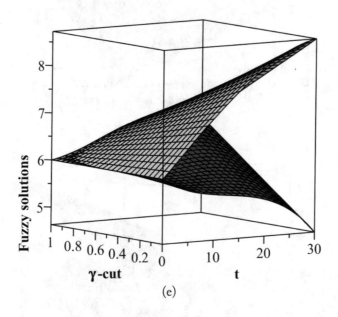

(e)

Figure 7.1: (*Continued.*) Fuzzy solution of fractional order HEV model for (e) $\tilde{P}(t; \gamma)$ where $t \in [0, 30]$ and $\gamma \in [0, 1]$ when $\alpha = 1$.

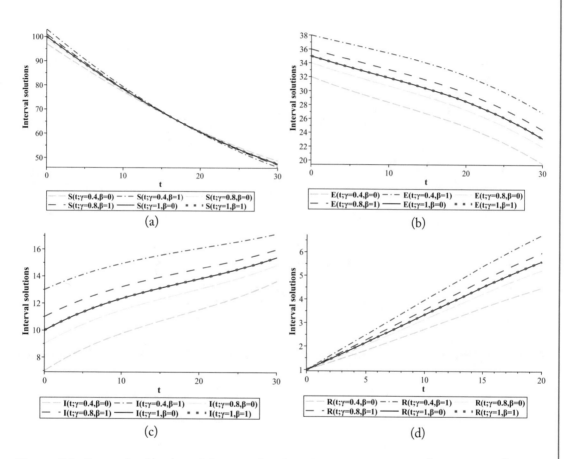

Figure 7.2: Interval solutions of fractional order HEV model for (a) $\tilde{S}(t;\gamma)$, (b) $\tilde{E}(t;\gamma)$, (c) $\tilde{I}(t;\gamma)$, and (d) $\tilde{R}(t;\gamma)$ and when $\alpha = 1$. (*Continues.*)

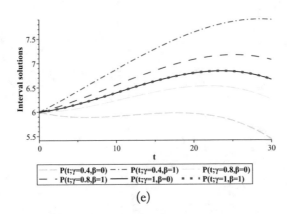

(e)

Figure 7.2: (*Continued.*) Interval solutions of fractional order HEV model for (e) $\tilde{P}\,(t\,;\gamma)$ when $\alpha = 1$.

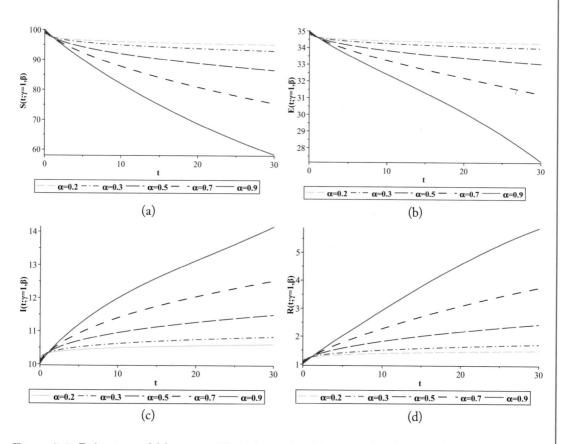

Figure 7.3: Behaviors of (a) susceptible individuals, (b) exposed individuals, (c) infected individuals, and (d) recovered individuals for different values of α when there is a change in time. (*Continues.*)

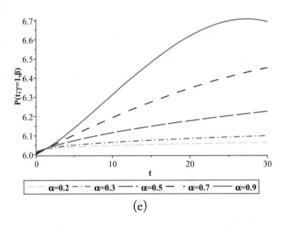

(e)

Figure 7.3: (*Continued.*) Behaviors of (a) susceptible individuals and (e) density of viral load for different values of α when there is a change in time.

7.5   REFERENCES

[1] K. Das, A. Agarwal, R. Andrew, G. G. Frösnerm, and P. Kar. Role of hepatitis E and other hepatotropic virus in aetiology of sporadic acute viral hepatitis: A hospital-based study from urban Delhi. *European Journal of Epidemiology*, 16:937–940, 2000. DOI: 10.1023/A:1011072015127. 105

[2] R. Viswanathan. Infectious hepatitis in Dehli (1955–1956): A critical study; epidemiology. *Indian Journal of Medical Research*, 45:1–30, 1957. 105

[3] WHO Report, 2009. http://www.who.int/countries/uga/en/ 105

[4] D. B. Rein, G. A. Stevens, J. Theaker, J. S. Wittenborn, and S. T. Wiersma. The global burden of hepatitis E virus genotypes 1 and 2 in 2005. *Hepatology*, 55:988–997, 2012. DOI: 10.1002/hep.25505. 105

[5] D. G. Prakasha, P. Veeresha, and H. M. Baskonus. Analysis of the dynamics of hepatitis E virus using the Atangana–Baleanu fractional derivative. *European Physical Journal Plus*, 134:241, 2019. DOI: 10.1140/epjp/i2019-12590-5. 105

[6] M. A. Khan, Z. Hammouch, and D. Baleanu. Modeling the dynamics of hepatitis E via the Caputo–Fabrizio derivative. *Mathematical Modelling of Natural Phenomena*, 14:311, 2019. DOI: 10.1051/mmnp/2018074. 106

[7] S. Ullah, M. A. Khan, and M. Farooq. A new fractional model for the dynamics of the hepatitis B virus using the Caputo–Fabrizio derivative. *European Physical Journal Plus*, 133:237, 2018. DOI: 10.1140/epjp/i2018-12072-4. 105

[8] E. Ahmed and H. A. El-Saka. On fractional-order models for hepatitis C. *Nonlinear Biomedical Physics*, 4:1–10, 2010. DOI: 10.1186/1753-4631-4-1. 105

[9] M. Andraud, M. Dumarest, R. Cariolet, B. Aylaj, E. Barnaud, F. Eono, N. Pavio, and N. Rose. Direct contact and environmental contaminations are responsible for HEV transmission in pigs. *Veterinary Research*, 44:0928–4249, 2013. DOI: 10.1186/1297-9716-44-102. 105

[10] G. N. Mercera and M. R. Siddiqui. Application of a hepatitis E transmission model to assess intervention strategies in a displaced persons camp in Uganda. *19th International Congress on Modelling and Simulation*, pp. 12–16, Perth, Australia, 2011. 105

[11] B. Nannyonga et al. The dynamics, causes and possible prevention of hepatitis E outbreaks. *PLoS One*, 7:e41135, 2012. DOI: 10.1371/journal.pone.0041135. 105

[12] H. Ren et al. The development of a combined mathematical model to forecast the incidence of hepatitis E in Shanghai, China. *BMC Infectious Diseases*, 13:421, 2013. DOI: 10.1186/1471-2334-13-421. 105

[13] R. M. Jena, S. Chakraverty, and D. Baleanu. On the solution of imprecisely defined nonlinear time-fractional dynamical model of marriage. *Mathematics*, 7:689–704, 2019. DOI: 10.3390/math7080689. 105

[14] R. M. Jena, S. Chakraverty, and D. Baleanu. On new solutions of time-fractional wave equations arising in Shallow water wave propagation. *Mathematics*, 7:722–734, 2019. DOI: 10.3390/math7080722.

[15] R. M. Jena and S. Chakraverty. Solving time-fractional Navier–Stokes equations using homotopy perturbation Elzaki transform. *SN Applied Sciences*, 1(1):16, 2019. DOI: 10.1007/s42452-018-0016-9.

[16] R. M. Jena and S. Chakraverty. Residual power series method for solving time-fractional model of vibration equation of large membranes. *Journal of Applied and Computational Mechanics*, 5:603–615, 2019. DOI: 10.22055/jacm.2018.26668.1347.

[17] R. M. Jena and S. Chakraverty. A new iterative method based solution for fractional Black–Scholes option pricing equations (BSOPE). *SN Applied Sciences*, 1:95–105, 2019. DOI: 10.1007/s42452-018-0106-8.

[18] R. M. Jena and S. Chakraverty. Analytical solution of Bagley–Torvik equations using Sumudu transformation method. *SN Applied Sciences*, 1(3):246, 2019. DOI: 10.1007/s42452-019-0259-0.

[19] R. M. Jena, S. Chakraverty, and S. K. Jena. Dynamic response analysis of fractionally damped beams subjected to external loads using homotopy analysis method. *Journal of Applied and Computational Mechanics*, 5:355–366, 2019. DOI: 10.22055/jacm.2019.27592.1419. 105

[20] S. Chakraverty, S. Tapaswini, and D. Behera. *Fuzzy Arbitrary Order System: Fuzzy Fractional Differential Equations and Applications*. John Wiley & Sons, 2016. DOI: 10.1002/9781119004233.

[21] S. Chakraverty, S. Tapaswini, and D. Behera. *Fuzzy Differential Equations and Applications for Engineers and Scientists*. Taylor & Francis Group, CRC Press, Boca Raton, FL, 2016. DOI: 10.1201/9781315372853.

[22] S. Chakraverty, D. M. Sahoo, and N. R. Mahato. *Concepts of Soft Computing: Fuzzy and ANN with Programming*. Springer, Singapore, 2019. DOI: 10.1007/978-981-13-7430-2.

CHAPTER 8

Fuzzy Time-Fractional SIRS-SI Malaria Disease Model

8.1 INTRODUCTION

Malaria is a mosquito-borne infectious illness caused by a parasite of the plasmodium group. The carrier of the plasmodium parasite is the female anopheles mosquito, which causes the destruction of red platelets in human beings and animals through bites. Reports from the WHO show that malaria is a significant hazard to human life. It may be transmitted through blood transfusion and sharing needles. Mathematical modeling of such types of infectious diseases is very important to understand the dynamical behavior of disease spreading and strategies to control it.

Recently, researchers have studied the mathematical models for malaria transmission. In this regard, the system of differential equations related to the transmission of malaria disease is studied by Agusto et al. [1]. They applied the optimal control theory to examine the optimal strategies for controlling the spread of malaria, using treatments like insecticide-sprayed nets and spray of mosquito insecticide. Abdullahi et al. [2] studied the possible outcomes of treatment on malaria transmission. Mandal et al. [3] developed mathematical modeling for malaria with the circumstance of individual-to-individual transmission through blood transfusions and malaria-infected women having pregnancy. Chiyaka et al. [4] proposed an improved mathematical modeling assuming that humans are in the recovered group and have a possibility of being susceptible. Rafikov et al. [5] suggested a strategy to control malaria using genetically altered mosquitoes. Yang [6] presented an approach for describing the malaria transmission associated with global warming and socioeconomic circumstances. Senthamarai et al. [7] applied the homotopy analysis method to observe the spreading of malaria in a SIRS-SI model. However, all these kinds of techniques and models have their own drawbacks because of the local nature of integer-order derivatives. Thus, fractional orders are recommended in mathematical modeling of biological and physical systems [8–20].

The main purpose of this chapter is to obtain the solution of the time-fractional SIRS-SI malaria disease model using FRDTM in an uncertain environment. In this model, initial conditions are considered as uncertain in terms of TFN.

8.2   MATHEMATICAL MODEL

It is important to understand biological problems with the help of developing mathematical models and analyzing their dynamical behaviors. In the present work, we have investigated the system of time-fractional SIRS-SI malaria disease model in the Caputo sense, which consists of three classes: Class I: Susceptible human S_h, Class II: Infected human I_h, Class III: Recovered human R_h. Further, the number of mosquitoes is classified into two classes: Class I: Susceptible mosquito S_m; Class II: Infected mosquito I_m. Let us assume that a newborn baby is transferred to the susceptible class with a rate α_h per unit of time. The susceptible-class humans may move into the infected class due to blood transfusion at a rate $\omega\gamma_1$ (where ω is the average number of blood transfusion between the susceptible and infected classes per unit of time, whereas γ_1 denotes the chance of disease transmission to susceptible human from infected human) or through an infected mosquito bite with a rate $\xi\gamma_2$ (where ξ indicates the average number of infected mosquito bites on susceptible humans per unit of time and γ_2 is the probability of disease transmission from infected mosquitoes to susceptible humans). The susceptible-group humans may move into the recovered group using vaccinations with a rate δ per unit of time. Persons in susceptible groups expire at a rate of η_h. A newborn baby can be infected by malaria due to hereditary with a rate μ per unit of time. Persons in the infected group can move to the recovered group due to the use of antimalarial drugs with a rate cv per unit of time (where c is the recovery rate of humans and v is the efficiency of anti-malarial drugs). The infected-group persons can die at a rate η_h and death due to malaria disease at a rate ε per unit of time. Persons in the recovered group may die at a rate η_h. Furthermore, newborn mosquitoes are transferred to the susceptible group at a constant rate α_m per unit of time. The susceptible-class mosquitoes may transfer into the infected class by biting infected persons with a rate of $E\gamma_3$ (where E stands for the average number of susceptible mosquito bites on infected humans per unit of time and γ_3 indicates the probability of disease transmission from infected people to susceptible mosquitoes) or can die with a fixed rate η_m per unit of time. b denotes the average rate of loss of immunity per capita per unit of time. The mosquitoes in susceptible and infected groups can expire due to the use of spraying at a rate σ per unit of time. The infected-class mosquitoes can die with the use of vaccines, anti-malarial drugs, and spraying at a rate η_m per unit of time. The governing equations for the time-fractional SIRS-SI malaria disease in the Caputo sense are given as [7, 18–20]

$$
\begin{cases}
\frac{d^\alpha S_h}{dt^\alpha} = \alpha_h + bR_h - (\omega\gamma_1 I_h + \xi\gamma_2 I_m) S_h - (\delta + \eta_h) S_h, \\[2mm]
\frac{d^\alpha I_h}{dt^\alpha} = \mu I_h + (\omega\gamma_1 I_h + \xi\gamma_2 I_m) S_h - (\eta_h + \varepsilon \mid cv) I_h, \\[2mm]
\frac{d^\alpha R_h}{dt^\alpha} = cv I_h - (\eta_h + b) R_h + \delta S_h, \\[2mm]
\frac{d^\alpha S_m}{dt^\alpha} = \alpha_m - (E\gamma_3 I_h + \eta_m + \sigma) S_m, \\[2mm]
\frac{d^\alpha I_m}{dt^\alpha} = E\gamma_3 I_h S_m - (\eta_m + \sigma) I_m,
\end{cases}
\tag{8.1}
$$

subject to initial conditions:

$$S_h\,(0) = 40, \quad I_h\,(0) = 2, \quad R_h\,(0) = 0, \quad S_m\,(0) = 500, \quad \text{and} \quad I_m\,(0) = 10. \tag{8.2}$$

8.3   MATHEMATICAL MODEL WITH FUZZY INITIAL CONDITIONS

Let us consider the following fuzzy time-fractional SIRS-SI malaria disease model as

$$\begin{cases} \frac{d^\alpha \tilde{S}_h}{dt^\alpha} = \alpha_h + b\tilde{R}_h - \left(\omega\gamma_1 \tilde{I}_h + \xi\gamma_2 \tilde{I}_m\right)\tilde{S}_h - (\delta + \eta_h)\,\tilde{S}_h, \\[2mm] \frac{d^\alpha \tilde{I}_h}{dt^\alpha} = \mu\tilde{I}_h + \left(\omega\gamma_1 \tilde{I}_h + \xi\gamma_2 \tilde{I}_m\right)\tilde{S}_h - (\eta_h + \varepsilon + cv)\,\tilde{I}_h, \\[2mm] \frac{d^\alpha \tilde{R}_h}{dt^\alpha} = cv\tilde{I}_h - (\eta_h + b)\,\tilde{R}_h + \delta\tilde{S}_h, \\[2mm] \frac{d^\alpha \tilde{S}_m}{dt^\alpha} = \alpha_m - \left(E\gamma_3 \tilde{I}_h + \eta_m + \sigma\right)\tilde{S}_m, \\[2mm] \frac{d^\alpha \tilde{I}_m}{dt^\alpha} = E\gamma_3 \tilde{I}_h\tilde{S}_m - (\eta_m + \sigma)\,\tilde{I}_m. \end{cases} \tag{8.3}$$

In this model, we have considered three initial conditions as the uncertain that are in terms of TFN, which may be written as follows:

$$\begin{aligned} &\tilde{S}_h\,(0) = (35, 40, 45), \quad I_h\,(0) = 2, \quad R_h\,(0) = 0, \\ &\tilde{S}_m\,(0) = (495, 500, 505), \quad \text{and} \quad \tilde{I}_m\,(0) = (5, 10, 15). \end{aligned} \tag{8.4}$$

Using the single parametric form, we may write Eq. (8.3) as

$$
\begin{cases}
\left[\frac{d^\alpha \underline{S}_h(t;\gamma)}{dt^\alpha}, \frac{d^\alpha \overline{S}_h(t;\gamma)}{dt^\alpha}\right] = \alpha_h + b\left[\underline{R}_h(t;\gamma), \overline{R}_h(t;\gamma)\right] \\
- \left\{\omega\gamma_1\left[\underline{I}_h(t;\gamma), \overline{I}_h(t;\gamma)\right] + \xi\gamma_2\left[\underline{I}_m(t;\gamma), \overline{I}_m(t;\gamma)\right]\right\}\left[\underline{S}_h(t;\gamma), \overline{S}_h(t;\gamma)\right] \\
- (\delta + \eta_h)\left[\underline{S}_h(t;\gamma), \overline{S}_h(t;\gamma)\right], \\
\left[\frac{d^\alpha \underline{I}_h(t;\gamma)}{dt^\alpha}, \frac{d^\alpha \overline{I}_h(t;\gamma)}{dt^\alpha}\right] = \mu\left[\underline{I}_h(t;\gamma), \overline{I}_h(t;\gamma)\right] \\
+ \left\{\omega\gamma_1\left[\underline{I}_h(t;\gamma), \overline{I}_h(t;\gamma)\right] + \xi\gamma_2\left[\underline{I}_m(t;\gamma), \overline{I}_m(t;\gamma)\right]\right\}\left[\underline{S}_h(t;\gamma), \overline{S}_h(t;\gamma)\right] \\
- (\eta_h + \varepsilon + cv)\left[\underline{I}_h(t;\gamma), \overline{I}_h(t;\gamma)\right], \\
\left[\frac{d^\alpha \underline{R}_h(t;\gamma)}{dt^\alpha}, \frac{d^\alpha \overline{R}_h(t;\gamma)}{dt^\alpha}\right] = cv\left[\underline{I}_h(t;\gamma), \overline{I}_h(t;\gamma)\right] - (\eta_h + b) \\
\left[\underline{R}_h(t;\gamma), \overline{R}_h(t;\gamma)\right] + \delta\left[\underline{S}_h(t;\gamma), \overline{S}_h(t;\gamma)\right], \\
\left[\frac{d^\alpha \underline{S}_m(t;\gamma)}{dt^\alpha}, \frac{d^\alpha \overline{S}_m(t;\gamma)}{dt^\alpha}\right] = \alpha_m - \left(E\gamma_3\left[\underline{I}_h(t;\gamma), \overline{I}_h(t;\gamma)\right] + \eta_m + \sigma\right) \\
\left[\underline{S}_m(t;\gamma), \overline{S}_m(t;\gamma)\right], \\
\left[\frac{d^\alpha \underline{I}_m(t;\gamma)}{dt^\alpha}, \frac{d^\alpha \overline{I}_m(t;\gamma)}{dt^\alpha}\right] = E\gamma_3\left[\underline{I}_h(t;\gamma), \overline{I}_h(t;\gamma)\right]\left[\underline{S}_m(t;\gamma), \overline{S}_m(t;\gamma)\right] \\
- (\eta_m + \sigma)\left[\underline{I}_m(t;\gamma), \overline{I}_m(t;\gamma)\right],
\end{cases}
\tag{8.5}
$$

subject to fuzzy initial conditions:

$$
\begin{aligned}
\left[\underline{S}_h(0;\gamma), \overline{S}_h(0;\gamma)\right] &= [5\gamma + 35, -5\gamma + 45], \quad I_h(0) = 2, \quad R_h(0) = 0, \\
\left[\underline{S}_m(0;\gamma), \overline{S}_m(0;\gamma)\right] &= [5\gamma + 495, -5\gamma + 505], \quad \text{and} \\
\left[\underline{I}_m(0;\gamma), \overline{I}_m(0;\gamma)\right] &= [5\gamma + 5, -5\gamma + 15],
\end{aligned}
\tag{8.6}
$$

where $\gamma \in [0, 1]$.

Next, by applying double parametric form (as discussed in the previous chapters), Eqs. (8.5) and (8.6) can be expressed as

$$
\begin{cases}
\left\{\beta\left(\frac{d^\alpha \overline{S}_h(t;\gamma)}{dt^\alpha} - \frac{d^\alpha \underline{S}_h(t;\gamma)}{dt^\alpha}\right) + \frac{d^\alpha \underline{S}_h(t;\gamma)}{dt^\alpha}\right\} = \alpha_h + b\left\{\beta\left(\overline{R}_h(t;\gamma) - \underline{R}_h(t;\gamma)\right)\right. \\
\left. + \underline{R}_h(t;\gamma)\right\} - \left\{\omega\gamma_1\left\{\beta\left(\overline{I}_h(t;\gamma) - \underline{I}_h(t;\gamma)\right) + \underline{I}_h(t;\gamma)\right\} + \xi\gamma_2\left\{\beta\left(\overline{I}_m(t;\gamma)\right.\right.\right. \\
\left.\left.\left. - \underline{I}_m(t;\gamma)\right) + \underline{I}_m(t;\gamma)\right\}\right\}\left\{\beta\left(\overline{S}_h(t;\gamma) - \underline{S}_h(t;\gamma)\right) + \underline{S}_h(t;\gamma)\right\} - (\delta + \eta_h) \\
\left\{\beta\left(\overline{S}_h(t;\gamma) - \underline{S}_h(t;\gamma)\right) + \underline{S}_h(t;\gamma)\right\}, \\
\left\{\beta\left(\frac{d^\alpha \overline{I}_h(t;\gamma)}{dt^\alpha} - \frac{d^\alpha \underline{I}_h(t;\gamma)}{dt^\alpha}\right) + \frac{d^\alpha \underline{I}_h(t;\gamma)}{dt^\alpha}\right\} = \mu\left\{\beta\left(\overline{I}_h(t;\gamma) - \underline{I}_h(t;\gamma)\right)\right. \\
\left. + \underline{I}_h(t;\gamma)\right\} + \left\{\omega\gamma_1\left\{\beta\left(\overline{I}_h(t;\gamma) - \underline{I}_h(t;\gamma)\right) + \underline{I}_h(t;\gamma)\right\} + \xi\gamma_2\left\{\beta\left(\overline{I}_m(t;\gamma) - \underline{I}_m(t;\gamma)\right)\right.\right. \\
\left.\left. + \underline{I}_m(t;\gamma)\right\}\right\}\left\{\beta\left(\overline{S}_h(t;\gamma) - \underline{S}_h(t;\gamma)\right) + \underline{S}_h(t;\gamma)\right\} - (\eta_h + \varepsilon + cv) \\
\left\{\beta\left(\overline{I}_h(t;\gamma) - \underline{I}_h(t;\gamma)\right) + \underline{I}_h(t;\gamma)\right\}, \left\{\beta\left(\frac{d^\alpha \overline{R}_h(t;\gamma)}{dt^\alpha} - \frac{d^\alpha \underline{R}_h(t;\gamma)}{dt^\alpha}\right) + \frac{d^\alpha \underline{R}_h(t;\gamma)}{dt^\alpha}\right\} \\
= cv\left\{\beta\left(\overline{I}_h(t;\gamma) - \underline{I}_h(t;\gamma)\right) + \underline{I}_h(t;\gamma)\right\} - (\eta_h + b) \\
\left\{\beta\left(\overline{R}_h(t;\gamma) - \underline{R}_h(t;\gamma)\right) + \underline{R}_h(t;\gamma)\right\} + \delta\left\{\beta\left(\overline{S}_h(t;\gamma) - \underline{S}_h(t;\gamma)\right) + \underline{S}_h(t;\gamma)\right\}, \\
\left\{\beta\left(\frac{d^\alpha \overline{S}_m(t;\gamma)}{dt^\alpha} - \frac{d^\alpha \underline{S}_m(t;\gamma)}{dt^\alpha}\right) + \frac{d^\alpha \underline{S}_m(t;\gamma)}{dt^\alpha}\right\} = \alpha_m \\
- \left(E\gamma_3\left\{\beta\left(\overline{I}_h(t;\gamma) - \underline{I}_h(t;\gamma)\right) + \underline{I}_h(t;\gamma)\right\} + \eta_m + \sigma\right)\left\{\beta\left(\overline{S}_m(t;\gamma) - \underline{S}_m(t;\gamma)\right)\right. \\
\left. + \underline{S}_m(t;\gamma)\right\}, \\
\left\{\beta\left(\frac{d^\alpha \overline{I}_m(t;\gamma)}{dt^\alpha} - \frac{d^\alpha \underline{I}_m(t;\gamma)}{dt^\alpha}\right) + \frac{d^\alpha \underline{I}_m(t;\gamma)}{dt^\alpha}\right\} = E\gamma_3\left\{\beta\left(\overline{I}_h(t;\gamma) - \underline{I}_h(t;\gamma)\right) + \underline{I}_h(t;\gamma)\right\} \\
\left\{\beta\left(\overline{S}_m(t;\gamma) - \underline{S}_m(t;\gamma)\right) + \underline{S}_m(t;\gamma)\right\} - (\eta_m + \sigma) \\
\left\{\beta\left(\overline{I}_m(t;\gamma) - \underline{I}_m(t;\gamma)\right) + \underline{I}_m(t;\gamma)\right\},
\end{cases}
$$

$$(8.7)$$

subject to initial conditions

$$
\begin{aligned}
&\left\{\beta\left(\overline{S}_h(0;\gamma) - \underline{S}_h(0;\gamma)\right) + \underline{S}_h(0;\gamma)\right\} = \beta(-10\gamma + 10) + 5\gamma + 35, \\
&I_h(0) = 2, \quad R_h(0) = 0, \\
&\left\{\beta\left(\overline{S}_m(0;\gamma) - \underline{S}_m(0;\gamma)\right) + \underline{S}_m(0;\gamma)\right\} = \beta(-10\gamma + 10) + 5\gamma + 495, \quad \text{and} \\
&\left\{\beta\left(\overline{I}_m(0;\gamma) - \underline{I}_m(0;\gamma)\right) + \underline{I}_m(0;\gamma)\right\} = \beta(-10\gamma + 10) + 5\gamma + 5.
\end{aligned}
$$

$$(8.8)$$

Let us denote

$$
\beta\left(\frac{d^\alpha \overline{S}_h(t;\gamma)}{dt^\alpha} - \frac{d^\alpha \underline{S}_h(t;\gamma)}{dt^\alpha}\right) + \frac{d^\alpha \underline{S}_h(t;\gamma)}{dt^\alpha} = \frac{d^\alpha \tilde{S}_h(t;\gamma,\beta)}{dt^\alpha},
$$

$$\beta \left(\frac{d^\alpha \overline{I}_h (t;\gamma)}{dt^\alpha} - \frac{d^\alpha \underline{I}_h (t;\gamma)}{dt^\alpha} \right) + \frac{d^\alpha \underline{I}_h (t;\gamma)}{dt^\alpha} = \frac{d^\alpha \tilde{I}_h (t;\gamma,\beta)}{dt^\alpha},$$

$$\beta \left(\frac{d^\alpha \overline{R}_h (t;\gamma)}{dt^\alpha} - \frac{d^\alpha \underline{R}_h (t;\gamma)}{dt^\alpha} \right) + \frac{d^\alpha \underline{R}_h (t;\gamma)}{dt^\alpha} = \frac{d^\alpha \tilde{R}_h (t;\gamma,\beta)}{dt^\alpha},$$

$$\beta \left(\frac{d^\alpha \overline{S}_m (t;\gamma)}{dt^\alpha} - \frac{d^\alpha \underline{S}_m (t;\gamma)}{dt^\alpha} \right) + \frac{d^\alpha \underline{S}_m (t;\gamma)}{dt^\alpha} = \frac{d^\alpha \tilde{S}_m (t;\gamma,\beta)}{dt^\alpha},$$

$$\beta \left(\frac{d^\alpha \overline{I}_m (t;\gamma)}{dt^\alpha} - \frac{d^\alpha \underline{I}_m (t;\gamma)}{dt^\alpha} \right) + \frac{d^\alpha \underline{I}_m (t;\gamma)}{dt^\alpha} = \frac{d^\alpha \tilde{I}_m (t;\gamma,\beta)}{dt^\alpha},$$

$$\beta \left(\overline{S}_h (t;\gamma) - \underline{S}_h (t;\gamma) \right) + \underline{S}_h (t;\gamma) = \tilde{S}_h (t;\gamma,\beta),$$

$$\beta \left(\overline{I}_h (t;\gamma) - \underline{I}_h (t;\gamma) \right) + \underline{I}_h (t;\gamma) = \tilde{I}_h (t;\gamma,\beta),$$

$$\beta \left(\overline{R}_h (t;\gamma) - \underline{R}_h (t;\gamma) \right) + \underline{R}_h (t;\gamma) = \tilde{R}_h (t;\gamma,\beta),$$

$$\beta \left(\overline{S}_m (t;\gamma) - \underline{S}_m (t;\gamma) \right) + \underline{S}_m (t;\gamma) = \tilde{S}_m (t;\gamma,\beta),$$

$$\beta \left(\overline{I}_m (t;\gamma) - \underline{I}_m (t;\gamma) \right) + \underline{I}_m (t;\gamma) = \tilde{I}_m (t;\gamma,\beta),$$

$$\beta \left(\overline{S}_h (0;\gamma) - \underline{S}_h (0;\gamma) \right) + \underline{S}_h (0;\gamma) = \tilde{S}_h (0;\gamma,\beta),$$

$$\beta \left(\overline{S}_m (0;\gamma) - \underline{S}_m (0;\gamma) \right) + \underline{S}_m (0;\gamma) = \tilde{S}_m (0;\gamma,\beta),$$

$$\beta \left(\overline{I}_m (0;\gamma) - \underline{I}_m (0;\gamma) \right) + \underline{I}_m (0;\gamma) = \tilde{I}_m (0;\gamma,\beta),$$

$$\beta (-10\gamma + 10) + 5\gamma + 35 = \psi_1,$$

$$\beta (-10\gamma + 10) + 5\gamma + 495 = \psi_2, \text{ and } \beta (-10\gamma + 10) + 5\gamma + 5 = \psi_3.$$

Plugging all the above equations into Eq. (8.7) and initial conditions, Eq. (8.8), we obtain the following expressions:

$$
\begin{cases}
\frac{d^\alpha \tilde{S}_h(t;\gamma,\beta)}{dt^\alpha} = \alpha_h + b\tilde{R}_h (t;\gamma,\beta) - \{\omega\gamma_1 \tilde{I}_h (t;\gamma,\beta) + \xi\gamma_2 \tilde{I}_m (t;\gamma,\beta)\} \tilde{S}_h (t;\gamma,\beta) \\
\quad - (\delta + \eta_h) \tilde{S}_h (t;\gamma,\beta), \\
\frac{d^\alpha \tilde{I}_h(t;\gamma,\beta)}{dt^\alpha} = \mu\tilde{I}_h (t;\gamma,\beta) + \{\omega\gamma_1 \tilde{I}_h (t;\gamma,\beta) + \xi\gamma_2 \tilde{I}_m (t;\gamma,\beta)\} \tilde{S}_h (t;\gamma,\beta) - \\
\quad (\eta_h + \varepsilon + cv) \tilde{I}_h (t;\gamma,\beta), \\
\frac{d^\alpha \tilde{R}_h(t;\gamma,\beta)}{dt^\alpha} = cv\tilde{I}_h (t;\gamma,\beta) - (\eta_h + b) \tilde{R}_h (t;\gamma,\beta) + \delta\tilde{S}_h (t;\gamma,\beta), \\
\frac{d^\alpha \tilde{S}_m(t;\gamma,\beta)}{dt^\alpha} = \alpha_m - \left(E\gamma_3 \tilde{I}_h (t;\gamma,\beta) + \eta_m + \sigma \right) \tilde{S}_m (t;\gamma,\beta), \\
\frac{d^\alpha \tilde{I}_m(t;\gamma,\beta)}{dt^\alpha} = E\gamma_3 \tilde{I}_h (t;\gamma,\beta) \tilde{S}_m (t;\gamma,\beta) - (\eta_m + \sigma) \tilde{I}_m (t;\gamma,\beta),
\end{cases}
\tag{8.9}
$$

with initial conditions:

$$\tilde{S}_h(0; \gamma, \beta) = \psi_1, I_h(0) = 2, R_h(0) = 0, \tilde{S}_m(0; \gamma, \beta) = \psi_2, \text{ and } \tilde{I}_h(0; \gamma, \beta) = \psi_3. \quad (8.10)$$

Now, by applying FRDTM to Eqs. (8.9), and (8.10) (fuzzy initial conditions), we have the following expressions, respectively:

$$
\begin{cases}
\tilde{S}_{h_{k+1}}(\gamma, \beta) = \frac{\Gamma(1+\alpha k)}{\Gamma(1+\alpha k+\alpha)}
\left\{
\begin{array}{l}
\alpha_h \delta(k) + b\tilde{R}_{h_k}(\gamma, \beta) - \\
\left\{
\begin{array}{l}
\omega\gamma_1 \sum_{i=0}^{k} \tilde{I}_{h_i}(\gamma, \beta) \tilde{S}_{h_{k-i}}(\gamma, \beta) \\
+ \xi\gamma_2 \sum_{i=0}^{k} \tilde{I}_{m_i}(\gamma, \beta) \tilde{S}_{h_{k-i}}(\gamma, \beta)
\end{array}
\right\} \\
- (\delta + \eta_h) \tilde{S}_{h_k}(\gamma, \beta)
\end{array}
\right\}, \\[3em]

\tilde{I}_{h_{k+1}}(\gamma, \beta) = \frac{\Gamma(1+\alpha k)}{\Gamma(1+\alpha k+\alpha)}
\left\{
\begin{array}{l}
\mu\tilde{I}_{h_k}(\gamma, \beta) +
\left\{
\begin{array}{l}
\omega\gamma_1 \sum_{i=0}^{k} \tilde{I}_{h_i}(\gamma, \beta) \tilde{S}_{h_{k-i}}(\gamma, \beta) + \\
\xi\gamma_2 \sum_{i=0}^{k} \tilde{I}_{m_i}(\gamma, \beta) \tilde{S}_{h_{k-i}}(\gamma, \beta)
\end{array}
\right\} \\
- (\eta_h + \varepsilon + cv) \tilde{I}_{h_k}(\gamma, \beta)
\end{array}
\right\}, \\[3em]

\tilde{R}_{h_{k+1}}(\gamma, \beta) = \frac{\Gamma(1+\alpha k)}{\Gamma(1+\alpha k+\alpha)} \left\{ cv\tilde{I}_{h_k}(\gamma, \beta) - (\eta_h + b)\tilde{R}_{h_k}(\gamma, \beta) + \delta\tilde{S}_{h_k}(\gamma, \beta) \right\}, \\[1.5em]

\tilde{S}_{m_{k+1}}(\gamma, \beta) = \frac{\Gamma(1+\alpha k)}{\Gamma(1+\alpha k+\alpha)}
\left\{
\begin{array}{l}
\alpha_m \delta(k) - \left(E\gamma_3 \sum_{i=0}^{k} \tilde{I}_{h_i}(\gamma, \beta) \tilde{S}_{m_{k-i}}(\gamma, \beta) \right) \\
+ (\eta_m + \sigma) \tilde{S}_{m_k}(\gamma, \beta)
\end{array}
\right\}, \\[2em]

\tilde{I}_{m_{k+1}}(\gamma, \beta) = \frac{\Gamma(1+\alpha k)}{\Gamma(1+\alpha k+\alpha)} \left\{ E\gamma_3 \sum_{i=0}^{k} \tilde{I}_{h_i}(\gamma, \beta) \tilde{S}_{m_{k-i}}(\gamma, \beta) - (\eta_m + \sigma) \tilde{I}_{m_k}(\gamma, \beta) \right\},
\end{cases}
$$

$$(8.11)$$

subject to transformed initial conditions:

$$\tilde{S}_{h_0}(\gamma, \beta) = \psi_1, \tilde{I}_{h_0}(\gamma, \beta) = 2, \tilde{R}_{h_0}(\gamma, \beta) = 0,$$
$$\tilde{S}_{m_0}(\gamma, \beta) = \psi_2, \text{ and } \tilde{I}_{h_0}(\gamma, \beta) = \psi_3, \quad (8.12)$$

where,

$$\delta(k) = \begin{cases} 1, k = 0, \\ 0, k \neq 0. \end{cases} \quad (8.13)$$

Substituting Eq. (8.12) into Eq. (8.11) for $k = 0, 1, 2, \ldots$, the following expressions are obtained successively:

$$\tilde{S}_{h_1}(\gamma, \beta) = \frac{1}{\Gamma(1+\alpha)} (0.027 - 0.05192\psi_1 - 0.00130\psi_1\psi_3),$$

$$\tilde{I}_{h_1}(\gamma, \beta) = \frac{1}{\Gamma(1+\alpha)} (-0.1519 + 0.00152\psi_1 + 0.00130\psi_1\psi_3),$$

$$\tilde{R}_{h_1}(\gamma, \beta) = \frac{1}{\Gamma(1+\alpha)} (0.06110 + 0.05\psi_1),$$

$$\tilde{S}_{m_1}(\gamma, \beta) = \frac{1}{\Gamma(1+\alpha)}(0.13 - 0.093168\psi_2),$$

$$\tilde{I}_{m_1}(\gamma, \beta) = \frac{1}{\Gamma(1+\alpha)}(0.003168\psi_2 - 0.09\psi_3),$$

$$\tilde{S}_{h_2}(\gamma, \beta) = \frac{\Gamma(1+\alpha)}{\Gamma(1+2\alpha)}$$

$$\begin{pmatrix} \frac{0.0611+0.05\psi_1}{730\Gamma(1+\alpha)} - \frac{1}{\Gamma(1+\alpha)}(0.05192(0.027 - 0.05192\psi_1 - 0.0013\psi_1\psi_3)) \\ -\frac{0.00076\psi_1}{\Gamma(1+\alpha)}(-0.1519 + 0.00152\psi_1 + 0.0013\psi_1\psi_3) - \frac{0.0013\psi_3}{\Gamma(1+\alpha)} \\ (0.027 - 0.05192\psi_1 - 0.0013\psi_1\psi_3) - \frac{0.0013\psi_1}{\Gamma(1+\alpha)}(0.003168\psi_2 - 0.09\psi_3) \end{pmatrix},$$

$$\tilde{I}_{h_2}(\gamma, \beta) = \frac{\Gamma(1+\alpha)}{\Gamma(1+2\alpha)}$$

$$\begin{pmatrix} \frac{-0.07595}{\Gamma(1+\alpha)}(-0.1519 + 0.00152\psi_1 + 0.0013\psi_1\psi_3) + \frac{0.00152}{\Gamma(1+\alpha)} \\ (0.027 - 0.05912\psi_1 - 0.0013\psi_1\psi_3) + \frac{0.00076\psi_1}{\Gamma(1+\alpha)} \\ (-0.1519 + 0.00152\psi_1 + 0.0013\psi_1\psi_3) + \frac{0.0013\psi_3}{\Gamma(1+\alpha)} \\ (0.027 - 0.05912\psi_1 - 0.0013\psi_1\psi_3) + \frac{0.0013\psi_1}{\Gamma(1+\alpha)}(0.003168\psi_2 - 0.09\psi_3) \end{pmatrix},$$

$$\tilde{R}_{h_2}(\gamma, \beta) = \frac{\Gamma(1+\alpha)}{\Gamma(1+2\alpha)}$$

$$\begin{pmatrix} \frac{0.03055}{\Gamma(1+\alpha)}(-0.1519 + 0.00152\psi_1 + 0.0013\psi_1\psi_3) - \frac{0.001769863014}{\Gamma(1+\alpha)} \\ (0.0611 + 0.05\psi_1) + \frac{0.05}{\Gamma(1+\alpha)}(0.027 - 0.05912\psi_1 - 0.0013\psi_1\psi_3) \end{pmatrix},$$

$$\tilde{S}_{m_2}(\gamma, \beta) = \frac{\Gamma(1+\alpha)}{\Gamma(1+2\alpha)} \begin{pmatrix} \frac{-0.093168}{\Gamma(1+\alpha)}(0.13 - 0.093168\psi_2) - \frac{0.001584\psi_2}{\Gamma(1+\alpha)} \\ (-0.1519 + 0.00152\psi_1 + 0.0013\psi_1\psi_3) \end{pmatrix},$$

$$\tilde{I}_{m_2}(\gamma, \beta) = \frac{\Gamma(1+\alpha)}{\Gamma(1+2\alpha)}$$

$$\begin{pmatrix} \frac{0.003168}{\Gamma(1+\alpha)}(0.13 - 0.093168\psi_2) - \frac{0.001584\psi_2}{\Gamma(1+\alpha)}\begin{pmatrix} -0.1519 + 0.00152\psi_1 \\ +0.0013\psi_1\psi_3 \end{pmatrix} \\ -\frac{0.09_1}{\Gamma(1+\alpha)}(0.003168\psi_2 - 0.09\psi_3) \end{pmatrix},$$

Continuing the procedure in this way, we can find all the values of $\{\tilde{S}_{h_k}\}_{k=0}^{\infty}$, $\{\tilde{I}_{h_k}\}_{k=0}^{\infty}$, $\{\tilde{R}_{h_k}\}_{k=0}^{\infty}$, $\{\tilde{S}_{m_k}\}_{k=0}^{\infty}$, and $\{\tilde{I}_{m_k}\}_{k=0}^{\infty}$. Using inverse differential transform to $\left\{\tilde{S}_{h_k}\right\}_{k=0}^{\infty}, \left\{\tilde{I}_{h_k}\right\}_{k=0}^{\infty}, \left\{\tilde{R}_{h_k}\right\}_{k=0}^{\infty}, \left\{\tilde{S}_{m_k}\right\}_{k=0}^{\infty}, \left\{\tilde{I}_{m_k}\right\}_{k=0}^{\infty}$ and substituting the values of $\psi_1, \psi_2,$ and ψ_3, we have the following nth-order approximate solutions:

$$
\begin{cases}
\tilde{S}_{h_n}(t; \gamma, \beta) = \sum_{k=0}^{n} \tilde{S}_{h_k}(\gamma, \beta) t^{\alpha k}, & \tilde{I}_{h_n}(t; \gamma, \beta) = \sum_{k=0}^{n} \tilde{I}_{h_k}(\gamma, \beta) t^{\alpha k}, \\
\tilde{R}_{h_n}(t; \gamma, \beta) = \sum_{k=0}^{n} \tilde{R}_{h_k}(\gamma, \beta) t^{\alpha k}, & \tilde{S}_{m_n}(t; \gamma, \beta) = \sum_{k=0}^{n} \tilde{S}_{m_k}(\gamma, \beta) t^{\alpha k}, \\
\tilde{I}_{m_n}(t; \gamma, \beta) = \sum_{k=0}^{n} \tilde{I}_{m_k}(\gamma, \beta) t^{\alpha k},
\end{cases}
\tag{8.14}
$$

we may write the exact solution of this model as

$$
\begin{cases}
\tilde{S}_h(t; \gamma, \beta) = \lim_{n \to \infty} \tilde{S}_{h_n}(t; \gamma, \beta), \tilde{I}_h(t; \gamma, \beta) = \lim_{n \to \infty} \tilde{I}_{h_n}(t; \gamma, \beta) \\
\tilde{R}_h(t; \gamma, \beta) = \lim_{n \to \infty} \tilde{R}_{h_n}(t; \gamma, \beta), \tilde{S}_m(t; \gamma, \beta) = \lim_{n \to \infty} \tilde{S}_{m_n}(t; \gamma, \beta), \\
\tilde{I}_m(t; \gamma, \beta) = \lim_{n \to \infty} \tilde{I}_{m_n}(t; \gamma, \beta).
\end{cases}
\tag{8.15}
$$

The lower- and upper-bound solutions for this model can be evaluated, respectively, by substituting $\beta = 0$ and $\beta = 1$ into Eq. (8.14). Mathematically, we may write the following equations:

$$
\begin{cases}
\tilde{S}_h(t; \gamma, 0) = \underline{S_h}(t; \gamma), \tilde{I}_h(t; \gamma, 0) = \underline{I_h}(t; \gamma), \tilde{R}_h(t; \gamma, 0) = \underline{R_h}(t; \gamma), \\
\tilde{S}_m(t; \gamma, 0) = \underline{S_m}(t; \gamma), \tilde{I}_m(t; \gamma, 0) = \underline{I_m}(t; \gamma), \\
\text{and} \\
\tilde{S}_h(t; \gamma, 1) = \overline{S_h}(t; \gamma), \tilde{I}_h(t; \gamma, 1) = \overline{I_h}(t; \gamma), \tilde{R}_h(t; \gamma, 1) = \overline{R_h}(t; \gamma), \\
\tilde{S}_m(t; \gamma, 1) = \overline{S_m}(t; \gamma), \tilde{I}_m(t; \gamma, 1) = \overline{I_m}(t; \gamma).
\end{cases}
\tag{8.16}
$$

Similarly, one may consider other involved parameters as well as initial conditions as the fuzzy number and may proceed as per the above procedure.

8.4   RESULTS AND DISCUSSION

This section deals with the numerical computation for the time-fractional SIRS-SI malaria disease model. The numerical results are computed by using FRDTM. Several numerical calculations have been performed by considering various values of the parameters involved in the titled model.

   In order to obtain the approximate solution of this model, we have taken the values of the parameters as $\eta_h = 0.0004, \eta_m = 0.04, \varepsilon = 0.05, \sigma = 0.05, \alpha_h = 0.027, \alpha_m = 0.13, \omega = 0.038,$ $\xi = 0.13, \gamma_1 = 0.02, \gamma_2 = 0.01, \gamma_3 = 0.072, b = 1/730, c = 0.611, E = 0.022, \mu = 0.005, \delta = 0.05,$ and $v = 0.25$, which are taken from various reliable sources [7, 18, 19]. All the numerical calculations and plots are incorporated by considering the second-order ($n = 2$) approximate solutions. In this study, the susceptible human (S_h), susceptible mosquito, and infected mosquito

Table 8.1: Lower- and upper-bound solutions of the model at $\alpha = 1$

$t \rightarrow$		0.3	0.7	1.0
$\gamma = 0.3$	$[\underline{S_h}, \overline{S_h}]$	[35.85070, 42.60904]	[34.99634, 41.44450]	[34.36410, 40.58866]
	$[\underline{I_h}, \overline{I_h}]$	[2.064274, 2.200382]	[2.152067, 2.458503]	[2.219482, 2.645303]
	$[\underline{R_h}, \overline{R_h}]$	[0.561074, 0.66484]	[1.294376, 1.532684]	[1.833256, 2.169598]
	$[\underline{S_m}, \overline{S_m}]$	[482.8475, 489.6377]	[465.2225, 471.6893]	[452.4375, 458.6289]
	$[\underline{I_m}, \overline{I_m}]$	[6.793299, 13.63013]	[7.174801, 13.85378]	[7.453755, 14.05911]
$\gamma = 0.7$	$[\underline{S_h}, \overline{S_h}]$	[37.78544, 40.68187]	[36.84714, 39.61063]	[36.15426, 38.82192]
	$[\underline{I_h}, \overline{I_h}]$	[2.099393, 2.157726]	[2.231234, 2.362569]	[2.3296026, 2.51210]
	$[\underline{R_h}, \overline{R_h}]$	[0.590734, 0.635207]	[1.362526, 1.464658]	[1.9294808, 2.07362]
	$[\underline{S_m}, \overline{S_m}]$	[484.7881, 487.6982]	[467.0729, 469.8444]	[454.21215, 456.865]
	$[\underline{I_m}, \overline{I_m}]$	[8.746172, 11.67624]	[9.080308, 11.94271]	[9.3353404, 12.1661]
$\gamma = 1.0$	$[\underline{S_h}, \overline{S_h}]$	[39.23450, 39.23450]	[38.23079, 38.23079]	[37.490724, 37.4907]
	$[\underline{I_h}, \overline{I_h}]$	[2.127712, 2.127712]	[2.295015, 2.295015]	[2.4182590, 2.41826]
	$[\underline{R_h}, \overline{R_h}]$	[0.612973, 0.612973]	[1.4136064, 1.41360]	[2.00158, 2.00158]
	$[\underline{S_m}, \overline{S_m}]$	[486.2432, 486.2432]	[468.45934, 468.459]	[455.5401, 455.5401]
	$[\underline{I_m}, \overline{I_m}]$	[10.21109, 10.21109]	[10.51088, 10.51088]	[10.74948, 10.74948]

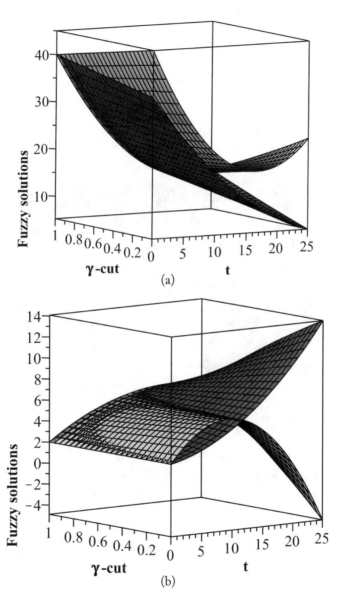

Figure 8.1: Uncertain solutions of Eq. (8.9) for (a) S_h and (b) I_h at $\alpha = 1$, $\gamma \in [0, 1]$ and $t \in [0, 25]$. (*Continues.*)

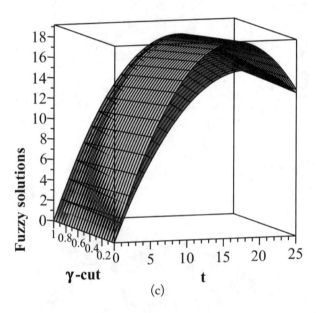

(c)

Figure 8.1: (*Continued.*) Uncertain solutions of Eq. (8.9) for (c) R_h at $\alpha = 1$, $\gamma \in [0, 1]$ and $t \in [0, 25]$.

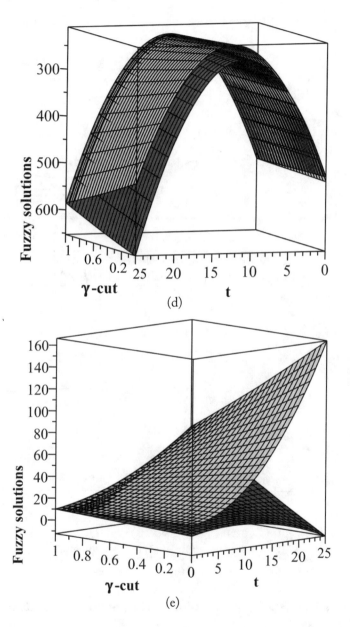

Figure 8.2: Uncertain solutions of Eq. (8.9) for (d) S_m and (e) I_m at $\alpha = 1, \gamma \in [0, 1]$ and $t \in [0, 25]$.

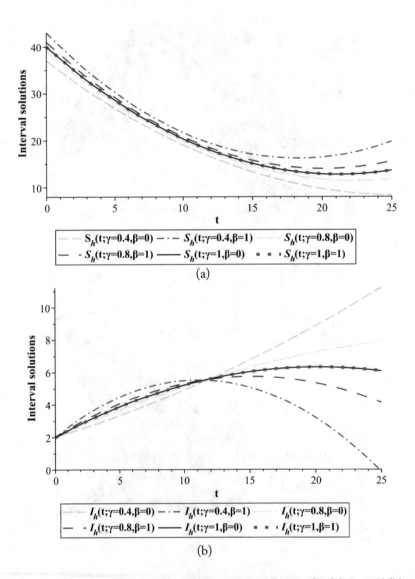

Figure 8.3: Interval solutions of Eq. (8.9) at various values of γ-cut for (a) S_h and (b) I_h at $\alpha = 1$, and $t \in [0, 25]$. (*Continues.*)

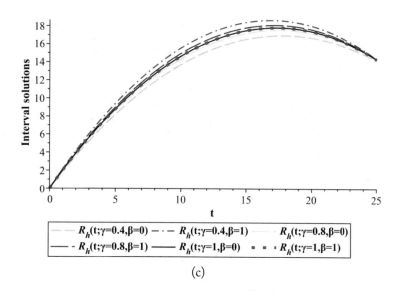

(c)

Figure 8.3: (*Continued.*) Interval solutions of Eq. (8.9) at various values of γ-cut for (c) R_h at $\alpha = 1$, and $t \in [0, 25]$.

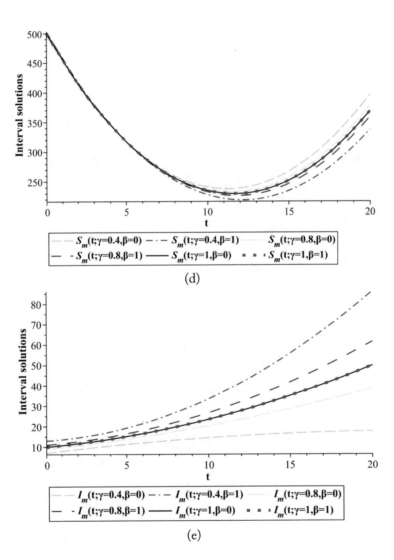

Figure 8.4: Interval solutions of Eq. (8.9) at various values of γ-cut for (d) S_m and (e) I_m at $\alpha = 1$, and $t \in [0, 20]$.

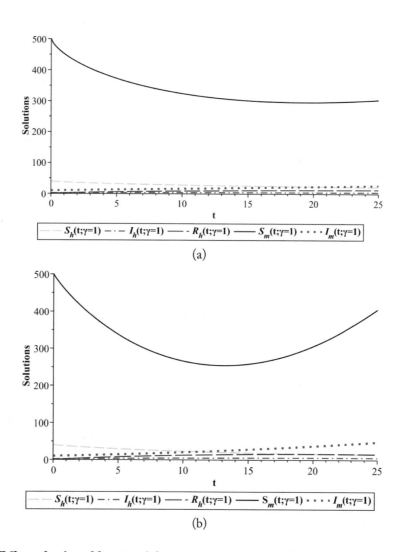

Figure 8.5: Effect of order of fractional derivatives on human and mosquito population at $\gamma = 1$, and $t \in [0, 25]$ when (a) $\alpha = 0.7$, and (b) $\alpha = 0.9$.

(I_m) at $t = 0$ are assumed to be the triangular fuzzy numbers. The triangular fuzzy lower-bound and upper-bound solutions of different groups of human population and mosquito populations of the titled model are depicted in Figs. 8.1 and 8.2 by changing the values of $t \in [0, 25]$ at $\alpha = 1$. Figures 8.3 and 8.4 demonstrate the influence of γ-cut on different classes of human and mosquito populations, respectively. One may conclude from Figs. 8.3 and 8.4 that the crisp result ($\gamma = 1$) is the central line and all the interval solutions are spread on both sides of the crisp results. It is worth mentioning that the present results at $\gamma = 1$ exactly match the solution plots of Putri et al. [19]. Figure 8.5 presents the solution plots of the titled model at different values of fractional order (α) ($= 0.7, 0.9$). It is noticed from Fig. 8.5 that the considered model considerably depends on the time-fractional order derivatives which assist to analyze the biological behavior of the proposed model. Table 8.1 includes the lower- and upper-bounds fuzzy solutions of the model at different values of t, γ, and $\alpha = 1$. From Table 8.1, it is clear that the lower bounds and the upper bounds are equal when $\gamma = 1$.

8.5   REFERENCES

[1] F. B. Agusto, N. Marcus, and K. O. Okosun. Application of optimal control to the epidemiology of malaria. *Electronic Journal of Differential Equations*, 81, 2012. 123

[2] M. B. Abdullahi, Y. A. Hasan, and F. A. Abdullah. A mathematical model of malaria and the effectiveness of drugs. *Applied Mathematical Sciences*, 7(62):3079–3095, 2013. DOI: 10.12988/ams.2013.13270. 123

[3] S. Mandal, R. R. Sarkar, and S. Sinha. Mathematical models of malaria—a review. *Malarian Journal*, 10:1–19, 2011. DOI: 10.1186/1475-2875-10-202. 123

[4] C. Chiyaka, J. M. Tchuenche, W. Garira, and S. Dube. A mathematical analysis of the effects of control strategies on the transmission dynamics of malaria. *Applied Mathematics and Computation*, 195:641–662, 2008. DOI: 10.1016/j.amc.2007.05.016. 123

[5] M. Rafikov, L. Bevilacqua, and A. P. P. Wyse. Optimal control strategy of malaria vector using genetically modified mosquitoes. *Journal of Theoretical Biology*, 258:418–425, 2009. DOI: 10.1016/j.jtbi.2008.08.006. 123

[6] H. M. Yang. A mathematical model for malaria transmission relating global warming and local socio economic conditions. *Revista de Saude Publica*, 35(3):224–231, 2001. DOI: 10.1590/s0034-89102001000300002. 123

[7] R. Senthamarai, S. Balamuralitharan, and A. Govindarajan. Application of homotopy analysis method in SIRS-SI model of malaria disease. *International Journal of Pure Applied Mathematics*, 113(12):239–248, 2017. 123, 124, 131

[8] R. M. Jena, S. Chakraverty, and D. Baleanu. On the solution of imprecisely defined nonlinear time-fractional dynamical model of marriage. *Mathematics*, 7:689–704, 2019. DOI: 10.3390/math7080689. 123

[9] R. M. Jena, S. Chakraverty, and D. Baleanu. On new solutions of time-fractional wave equations arising in Shallow water wave propagation. *Mathematics*, 7:722–734, 2019. DOI: 10.3390/math7080722.

[10] R. M. Jena and S. Chakraverty. Solving time-fractional Navier–Stokes equations using homotopy perturbation Elzaki transform. *SN Applied Sciences*, 1(1):16, 2019. DOI: 10.1007/s42452-018-0016-9.

[11] R. M. Jena and S. Chakraverty. Residual power series method for solving time-fractional model of vibration equation of large membranes. *Journal of Applied and Computational Mechanics*, 5:603–615, 2019. DOI: 10.22055/jacm.2018.26668.1347.

[12] R. M. Jena and S. Chakraverty. A new iterative method based solution for fractional Black–Scholes option pricing equations (BSOPE). *SN Applied Sciences*, 1:95–105, 2019. DOI: 10.1007/s42452-018-0106-8.

[13] R. M. Jena and S. Chakraverty. Analytical solution of Bagley–Torvik equations using Sumudu transformation method. *SN Applied Sciences*, 1(3):246, 2019. DOI: 10.1007/s42452-019-0259-0.

[14] R. M. Jena, S. Chakraverty, and S. K. Jena. Dynamic response analysis of fractionally damped beams subjected to external loads using homotopy analysis method. *Journal of Applied and Computational Mechanics*, 5:355–366, 2019. DOI: 10.22055/jacm.2019.27592.1419.

[15] S. Chakraverty, S. Tapaswini, and D. Behera. *Fuzzy Arbitrary Order System: Fuzzy Fractional Differential Equations and Applications*. John Wiley & Sons, 2016. DOI: 10.1002/9781119004233.

[16] S. Chakraverty, S. Tapaswini, and D. Behera. *Fuzzy Differential Equations and Applications for Engineers and Scientists*. Taylor & Francis Group, CRC Press, Boca Raton, FL, 2016. DOI: 10.1201/9781315372853.

[17] S. Chakraverty, D. M. Sahoo, and N. R. Mahato. *Concepts of Soft Computing: Fuzzy and ANN with Programming*. Springer, Singapore, 2019. DOI: 10.1007/978-981-13-7430-2.

[18] D. Kumar, J. Singh, M. A. Qurashi, and D. Baleanu. A new fractional SIRS-SI malaria disease model with application of vaccines, antimalarial drugs, and spraying. *Advances in Difference Equations*, 278, 2019. DOI: 10.1186/s13662-019-2199-9. 124, 131

[19] R. G. Putri, Jaharuddin, and T. Bakhtiar. Sirs-Si model of malaria disease with application of vaccines, anti-malarial drugs, and spraying. *IOSR Journal of Mathematics (IOSR-JM)*, 10(5):66–72, 2014. DOI: 10.9790/5728-10526672. 131, 140

[20] A. Ahmad, M. Farman, M. O. Ahmad, N. Raza, and M. Abdullah. Dynamical behavior of SIR epidemic model with non-integer time-fractional derivatives: A mathematical analysis. *International Journal of Advanced and Applied Sciences*, 5(1):123–129, 2018. DOI: 10.21833/ijaas.2018.01.016. 123, 124

Authors' Biographies

SNEHASHISH CHAKRAVERTY

Dr. Snehashish Chakraverty works in the Department of Mathematics (Applied Mathematics Group), National Institute of Technology Rourkela, Odisha, as a Senior (Higher Administrative Grade) Professor and is also the Dean of Student Welfare of the institute since November 2019. He received his Ph.D. from IIT Roorkee in 1992. Then he did post-doctoral research at ISVR, University of Southampton, U.K., and at Concordia University, Canada. He was a visiting professor at Concordia and McGill Universities, Canada, and University of Johannesburg, South Africa. Prof. Chakraverty has authored 17 books and published approximately 345 research papers in journals and conferences. He was the President of the Section of Mathematical Sciences (including Statistics) of Indian Science Congress (2015-2016) and was the Vice President—Orissa Mathematical Society (2011-2013). Prof. Chakraverty is a recipient of prestigious awards viz. INSA International Bilateral Exchange Program, Platinum Jubilee ISCA Lecture, CSIR Young Scientist, BOYSCAST, UCOST Young Scientist, Golden Jubilee CBRI Director's Award, Roorkee University gold Medals and more. He has undertaken 17 research projects as Principal Investigator funded by different agencies totaling about Rs.1.6 crores. Prof. Chakraverty is the Chief Editor of *International Journal of Fuzzy Computation and Modelling (IJFCM)*, Inderscience Publisher, Switzerland (http://www.inderscience.com/ijfcm), Associate Editor of *Computational Methods in Structural Engineering, Frontiers in Built Environment*, and an Editorial Board member of Springer Nature Applied Sciences, IGI Research Insights Books, Springer Book Series of Modeling and Optimization in Science and Technologies, Coupled Systems Mechanics (Techno Press), Curved and Layered Structures (De Gruyter), *Journal of Composites Science (MDPI)*, Engineering Research Express (IOP), *Applications and Applied Mathematics: An International Journal*, and Computational Engineering and Physical Modeling (Pouyan Press).

His present research area includes Differential Equations (Ordinary, Partial, and Fractional), Numerical Analysis and Computational Methods, Structural Dynamics (FGM, Nano), and Fluid Dynamics, Mathematical Modeling and Uncertainty Modeling, and Soft Computing and Machine Intelligence (Artificial Neural Network, Fuzzy, Interval and Affine Computations).

RAJARAMA MOHAN JENA

Rajarama Mohan Jena is currently working as a Senior Research Fellow at the Department of Mathematics, National Institute of Technology Rourkela, India. He is an INSPIRE (Innovation in Science Pursuit for Inspired Research) fellow of the Department of Science of Technology, Ministry of Science and Technology, Government of India, and doing his research under this fellowship. Rajarama does research in Fractional Dynamical Systems, Applied Mathematics, Computational Methods, Numerical Analysis, Partial Differential Equations, Fractional Differential Equations, Uncertainty Modeling, and Soft Computing, and so on. He has published 12 research papers in journals, two conference papers, and one book chapter. Rajarama has also served as a reviewer for various international journals. He has been continuing collaborative works with renowned researchers from Turkey, Canada, Iran, Egypt, Nigeria, United Kingdom, and other countries.

SUBRAT KUMAR JENA

Subrat Kumar Jena is currently working as a Research Fellow at the Department of Mathematics, National Institute of Technology Rourkela, India. He is also working in a Defence Research and Development Organisation (DRDO)-sponsored project entitled "Vibrations of Functionally Graded Nanostructural Members" in collaboration with Defence Metallurgical Research Laboratory (DMRL) Lab, Hyderabad. Subrat does research in Structural Dynamics, Nano Vibration, Applied Mathematics, Computational Methods, and Numerical Analysis, Uncertainty Modeling, Soft Computing, and other areas. He has published 16 research papers (till date) in journals, 5 conference papers, and 5 book chapters. He has also served as the reviewer for various international journals. He has been continuing collaborative research works with renowned researchers from Italy, Estonia, Turkey, Iran, Poland, and other countries.